A Roadmap for Transformative Science Teacher Leadership

This book is a comprehensive guide to an effective Science Education Fellowship (SEF) program. Spanning more than ten years and involving hundreds of teachers, District Science Coordinators, and university faculty, the Wipro SEF program has empowered teachers to become leaders who drive meaningful, sustainable change in their schools and districts without leaving the classroom.

Offering an in-depth look at the SEF program's structure, from its foundation in teacher leadership development to its innovative adaptations across seven universities and 35 school districts; the book presents a roadmap for implementing similar programs in other school districts, targeting teacher retention, teacher development, and fostering student growth. Readers will find detailed explanations of key program components, and the vital roles of district science coordinators and higher education institutions. Through a mix of theoretical insights, practical strategies, and testimonials from program participants, the book provides a comprehensive model for educators, administrators, and university leaders who aspire to replicate or adapt the SEF program in their own contexts.

Ideal for both educators and school administrators, this book will allow you to gain valuable insights into building and sustaining a program that empowers teacher-leaders, drives district-wide transformation, and ultimately improves student outcomes in science education.

Arthur Eisenkraft is a Distinguished Professor of Science Education, Professor of Physics, and Director of the Center of Science and Math in Context (COSMIC) at the University of Massachusetts Boston. He is a past president of the National Science Teaching Association and past chair of the Science Academic Advisory Committee of the College Board.

A Roadmap for Transformative Science Teacher Leadership

Building Meaningful Professional Development in Districts

Arthur Eisenkraft

Routledge
Taylor & Francis Group

NEW YORK AND LONDON

For Kaila

Contents

5 The District Science Coordinator   83

Arthur Eisenkraft and Larry R. Plank

6 The role of the Institution of Higher Education (IHE)   99

Arthur Eisenkraft

7 Innovation Phase of the Science Education Fellowship

Arthur Eisenkraft

8 Conclusion and the path forward

Arthur Eisenkraft

About the author

Dr. Arthur Eisenkraft

Dr. Arthur Eisenkraft is a distinguished leader in science education with a career spanning more than five decades. A product of the New York City public schools, he earned a BS in Physics from SUNY Stony Brook, an MA from the same institution, and a Ph.D. in Science Education from New York University, Dr. Eisenkraft has dedicated his life to advancing science education through teaching, curriculum development, and leadership.

Currently, Dr. Eisenkraft is a Distinguished Professor of Science Education and Professor of Physics at the University of Massachusetts Boston, where he also serves as the Director of the Center of Science and Math in Context (COSMIC). Under his leadership, COSMIC has become a hub for innovative science and math education projects, including the Wipro Science Education Fellowship (SEF), a program that has positively impacted thousands of students and hundreds of teachers across the United States.

Before joining UMass Boston, Dr. Eisenkraft spent two decades as a science coordinator and physics teacher at Bedford Public Schools in New York, where he influenced the next generation of scientists and educators. His career began as a Peace Corps volunteer in Nepal, and since then, he has taught in various capacities, including positions at Briarcliff High School and Ward Melville High School.

Dr. Eisenkraft's contributions to science education extend beyond the classroom. He has held leadership roles in numerous national and international organizations, including serving as President of the National Science Teaching Association (NSTA) and as chair of the Science Academic Advisory Committee of the College Board. He has also been a pivotal figure in the development of science curricula, such as the widely adopted Active Physics and Active Chemistry programs introducing high-quality project-based science to all students. He also chairs (and co-created) the Toshiba/NSTA ExploraVision Awards, involving 15,000 students annually for more than 30 years. He inaugurated the United States involvement in the International Physics Olympiad (with Jack M. Wilson) and served as the Academic Director of the team and then as Executive Director of the XXIV International Physics Olympiad involving the top pre-college physics students from 40 nations.

His achievements have earned him numerous accolades, including the National Science Board Public Service Award, the Robert H. Carleton Award from NSTA, the Robert A. Millikan Medal from the American Association of Physics Teachers (AAPT), the Presidential Award for Excellence in Mathematics and Science Teaching, and the Disney Corporation's Science Teacher of the Year. He is also a Fellow of the American Association for the Advancement of Science (AAAS) and the AAPT.. He holds a patent for a laser vision testing system and was awarded an honorary doctorate from Rensselaer Polytechnic Institute.

Dr. Eisenkraft is a prolific author and editor, contributing to several influential books and numerous articles in the field of science education. His work has consistently focused on improving science teaching and learning, with a particular emphasis on equity and access to quality education for all students.

As the overseer of the Wipro SEF program across seven universities and 35 school districts, Dr. Eisenkraft has been instrumental in fostering a culture of teacher leadership that drives sustainable district transformation. His passion for science education and his unwavering commitment to excellence continue to inspire educators and students alike, making him a true leader in the field.

Acknowledgments

It is with deep gratitude that I acknowledge the many individuals who have contributed to the success of the Wipro Science Education Fellowship (SEF) program over the past decade. This program has been a collective and collaborative effort, and it is only through the dedication, passion, and hard work of countless educators and partners that we have been able to make a meaningful impact on science education across the country.

First and foremost, I extend my heartfelt thanks to Anurag Behar, presently the CEO of the Azim Premji Foundation, who represented Wipro, our corporate funder. From the very beginning, Anurag was not just a provider of financial support but a true thought partner in the evolution of the program. His insight, commitment, and belief in the transformative power of teacher leadership have been invaluable. Wipro has set a powerful example of how a company can give back to the communities where its employees live and work by supporting teachers and helping school districts flourish. By way of introduction to many readers, Wipro Limited is a leading technology services and consulting company focused on building innovative solutions that address clients' most complex digital transformation needs. With more than 230,000 employees and business partners across 65 countries, they deliver on the promise of helping their customers, colleagues, and communities thrive in an ever-changing world.

My thanks extend to others at the Azim Premji Foundation and Wipro who have welcomed my presence on annual visits to India and allowed me to be a small part of their major effort toward improving education in India. This includes Dileep Ranjekar, Founder CEO, Azim Premji Foundation; PS Narayan, global head of sustainability and social initiatives, Wipro; Avinash Kumar, Associate Vice President at Wipro Digital Operations and Platform; and S. Giridhar, the first Registrar and COO of the Azim Premji University. It goes without saying that we thank Azim Premji and his generous philanthropy.

To the hundreds of Fellows who have participated in the SEF program, your dedication to your students, your school districts, and the pursuit of excellence in science education is nothing short of inspiring. Your hard work, your willingness to embrace new challenges, and your commitment to continuous growth have been the heart and soul of this program. Watching you present your work over the years has been a profound experience—your pride, expertise, and unwavering commitment to your craft are always evident. I often find myself wishing that those who criticize America's schools could see what you do each and every day; I am certain they would walk away with the respect and admiration you so richly deserve.

I also want to extend a special thank you to the District Science Coordinators who have been instrumental in guiding and supporting our Fellows. Your leadership has helped to ensure that the program aligns with district goals and that the impact of the SEF program is felt across entire school systems. Your tireless efforts to support teachers and improve science education are deeply appreciated. It has been a privilege to work with you over these years and to observe the learning community that you have developed amongst yourselves. As a science coordinator earlier in my career, I know all you have to juggle and I hope that the chapter describing your role in the Wipro SEF program somehow captures all of the talents you bring to the table.

To the professors and staff at the universities involved in the SEF program, thank you for your partnership, your expertise, and your unwavering commitment to excellence in education. The relationships that have developed between universities and school districts have been a cornerstone of this program's success. Your work has not only supported the professional growth of individual teachers but has also helped to build stronger, more resilient school systems.

A special word of thanks for Anthea Gabriel at the University of Massachusetts Boston. Anthea has been the steady heartbeat of the Wipro SEF and countless other projects at COSMIC over the years. While her formal title of Finance and Operations Manager may suggest a focus on logistics and numbers, it hardly captures the breadth of her contributions or the profound impact she has had on our work. Anthea's calm and composed demeanor has been a beacon of stability in what can often be a hectic environment. Her ability to navigate challenges with grace and efficiency has been instrumental in ensuring the success of our initiatives. The university is incredibly fortunate to have someone of her caliber, whose dedication goes beyond mere job descriptions. On a personal note, I have been deeply grateful for her support and wisdom, both professionally and personally. Anthea's presence has been invaluable, and her contributions have enriched our work and the lives of everyone at COSMIC.

Finally, I want to acknowledge the countless others – administrators, support staff, and community members – who have played a role in this journey. Your contributions, though perhaps less visible, have been equally important in creating the conditions for success.

Throughout the years of this program, I have had the privilege of witnessing the incredible work of teacher-leaders firsthand. Their passion, their creativity, and their commitment to their students have been a constant source of inspiration. This book is a testament to their efforts and to the belief that when we invest in our teachers, we invest in the future of our students and our communities.

Thank you all for being part of this journey. Your work has made a difference, and I am deeply grateful for your contributions. Together, we have built something truly special, and I look forward to seeing the continued impact of the SEF program in the years to come.

As we transition from the heartfelt thanks extended to those who made the Wipro SEF program a success, it is important to acknowledge the many individuals who played a crucial role in bringing this book to life. This book is more than just a collection of words; it is a reflection of a decade's worth of dedication, learning, and collaborative effort. It is intended to serve as a resource for teachers, school districts, and universities who wish to build on our experiences, whether by replicating the SEF program or by creating their own initiatives to foster district transformation through teacher leadership.

First, I want to express my deepest appreciation to the Fellows who generously shared their insights and experiences, whether through quotes specifically provided for this book or through the thoughtful reflections gathered during evaluation surveys over the years. Your voices have been integral in conveying the true impact of the SEF program. Your words bring authenticity and richness to these pages, making the lessons we've learned tangible and relatable to others who may follow in your footsteps.

A special thank you also goes to those who reviewed the manuscript at various stages. Your feedback was invaluable in refining the narrative and ensuring that the book effectively captures the essence of the SEF program. Your careful attention to detail, constructive critiques, and encouragement helped shape this book into a resource that will hopefully inspire and guide others in the field of education. This includes all of the university folk who reviewed early drafts and helped coax the project along. I particularly note the contributions and feedback at early and final stages of the book by our Wipro SEF Leadership Team – Tammy Wu Moriarty, Ratna Narayan, Meghan Marrero, Amanda Gunning, David Rosengrant, Allan Friedman, Linda Godwin, Meera Chandrasekhar, Mika Munakata, Emily Klein, Eric Weiss, Preetha Krishnan, Monica Taylor, Janet Carlson, Colette Killian, and Kristen Napolitano.

Anne Gurnee, our evaluator, helped lead an early session that led to an outline of the book and provided summaries of our data collections over the years that help ground our assertions of the success of Wipro SEF. Anne has taken on a leadership role in Wipro SEF as she guides our evaluation efforts. Marilyn Decker, Pam Pelletier, and Barnett Berry read all later drafts of each chapter. Marilyn and Pam have been with the program since its inception and their wisdom pervades both the program and the book. Barnett has extensive experience with programs supporting teacher efforts and awareness of the literature and I appreciate him finding the time to help us in this endeavor.

It is also important to recognize that while all university partners contributed to this book, some individuals took on central roles in developing specific chapters, ensuring that the story of the SEF program was told with clarity and coherence. These contributions were instrumental in weaving together the various elements of the program into a comprehensive narrative that highlights the program's successes and challenges.

At the University of Massachusetts Boston, Eric Weiss deserves a special thank you. He took volumes of our program notes, evaluations, reports and early drafts and made them coherent. The chapter on teacher leadership was made possible by the collaborative efforts of Amanda Gunning and Meghan Marrero. In addition, Brooke Whitworth and Jennifer Bateman reviewed the chapter and added some important research perspectives. The chapter on the Collaborative Coaching and Learning in Science model also benefitted from Amanda and Meghan's framing of different aspects of the program. The initial chapter on the role of District Science Coordinators was structured by Larry Plank. Additions to this chapter by Marilyn Decker and Pam Pelletier helped enormously. The chapter on the role of the Institutions of Higher Education benefitted from the contributions of Tammy Wu Moriarty, Meera Chandrasekhar, and Linda Godwin.

It is also crucial to acknowledge the behind-the-scenes work of those who helped compile, organize, and edit the contributions from various sources. Without their dedication and attention to detail, this book would not have come together as seamlessly as it did. From the initial concept to the final draft, their efforts ensured that the voices of the many contributors were harmonized into a cohesive whole.

Finally, I want to express my deepest gratitude to all the contributors whose names may not be explicitly mentioned here but whose efforts were no less significant. Whether you participated in discussions, shared your experiences, provided logistical support, or simply offered words of encouragement along the way, your contributions were vital to the creation of this book.

Together, we have created a resource that will serve as a beacon for educators, school districts, and universities striving to improve science education and foster teacher leadership. This book is a testament to the power of collaboration, the importance of shared knowledge, and the enduring impact of a community dedicated to educational excellence.

Thank you all for your hard work, your commitment, and your belief in the transformative power of education. It is our hope that this book will inspire others as much as it has inspired us, and that it will serve as a guide for those embarking on their own journeys toward educational innovation and excellence.

Wipro Science Education Fellowship Leadership Team

Dr. Janet Carlson is Associate Professor (Research) of Education at Stanford University and Faculty Director of CSET. Her research interests include the impact of educative curriculum materials and transformative professional development on science teaching and learning. She began her career as a middle and high school science teacher and has spent the past 20 years working in science education developing curriculum, leading professional development, and conducting research. Dr. Carlson received a BA in Environmental Biology from Carleton College, an MS in Curriculum and Instruction from Kansas State University, and a Ph.D. in Instruction and Curriculum (science education) from the University of Colorado.

Dr. Meera Chandrasekhar received a Ph.D. in Physics from Brown University in 1975. After a post-doctoral fellowship at the Max-Planck Institut in Stuttgart, Germany, she joined the faculty of the Physics department at the University of Missouri in Columbia, MO. Her research focused on optical studies of semiconductors and superconductors. In parallel, she pursued her interest in the education of K-12 students and teachers. She has received several awards for her work, including the A.P. Sloan Fellowship, the PAESMEM award from the National Science Foundation (NSF), and the Robert Foster Cherry Award for Teaching from Baylor University. She is currently Curators' Professor Emerita at the University of Missouri.

Marilyn Decker has a BS in physics and an MA in physics education from SUNY Stony Brook. She spent her career focusing on science teaching and science teachers. She served as the Director of STEM for the MA Department of Elementary and Secondary Education and has been a K-12 Science Director for several districts including Keene,

NH; Milton, MA; and Boston Public Schools. She was also Director of Analytical and Applied Sciences for the Jefferson County Public Schools, KY. She has been responsible for the development and implementation of the curriculum frameworks in science, technology, engineering, mathematics and computer science.

Dr. Arthur Eisenkraft is Distinguished Professor of Science Education, Professor of Physics and Director of the Center of Science and Math in Context (COSMIC) at the University of Massachusetts Boston. He is past president of the National Science Teaching Association and past chair of the Science Academic Advisory Committee of the College Board. He leads Active Physics and Active Chemistry introducing high quality project-based science to students. He also chairs (and co-created) the Toshiba/NSTA Explora-Vision Awards, involving 15,000 K-12 students annually for over 30 years.

Dr. Allan Feldman is Emeritus Professor of Science Education at the University of South Florida and the University of Massachusetts. He studies how people learn to do scientific research, and how in-service science teachers learn through action research. His books include *Teachers investigate their work: An introduction to action research across the professions* (2018), *Educating Science Teachers for Sustainability* (2015), and *Dialogic Collaborative Action Research in Science Education* (2023). Before receiving his doctorate, he taught middle and high school science for 17 years in public and private schools in New York, New Jersey, and Pennsylvania.

Dr. Linda Godwin, Professor Emerita of Physics and Astronomy at the University of Missouri (MU), joined the Missouri Wipro team in early 2023. Prior to joining the university as a professor in 2011, she spent 30 years at NASA Johnson Space Center where she was a veteran of four space shuttle flights. Her assignments at NASA included working in development of the future Exploration Program and as one of the original developers with NASA Headquarters education office on the Liftoff to Learning Series of products for K-12 using videos/graphics/teaching resources from space shuttle missions. At MU, she taught physics and astronomy classes and worked with students in undergraduate projects in astrophysics. She was a Co-Principal Investigator on a PhysTEC grant to encourage and improve the education of future physics teachers.

Dr. Amanda Gunning is Professor of Science Education at Mercy University where she teaches both content and methods courses for K-12 science and STEM teaching. Gunning is the Principal Investigator of the NSF-funded STEM Master Teacher Fellowship and co-directs the Greater NY Wipro Science Education Fellowship. Both provide research-driven PD for K-12 teacher fellows. Gunning presents her research on science teacher self-efficacy; STEM teacher development; vertical learning communities for teacher professional development; and family STEM learning at international conferences every year since 2009, and is published. She is the Co-Director and Co-Founder of Mercy University's Center for STEM Education.

Anne Gurnee is the founder and president of Anne Gurnee Consulting, LLC. Anne has worked in, with, and for businesses, formal and informal educational organizations (especially those focused on STEM), and non-profits for more than 30 years. With each project, she strives to bring her wealth of experience plus a passion for reflection through data to improve organizations and programs to make them stronger, more relevant, and better able to serve their audiences.

Colette Killian is Assistant Director of budgets and grants at Montclair State University. She has worked at Professional Resources in Science and Math (PRISM) housed at the Bristol-Myers Squibb Center for Science Teaching and Learning for more than 21 years. She works with multiple grants including the Wipro SEF grant, Computer Science NJ DOE grants, the NSF Secondary STEM Teacher Education Scholarship Program, BMS Block grants, NJ STEM Innovation grant, and multiple Research on Youth Thriving and Evaluation (RYTE) Institute grants. She assists with creating and planning grant-funded STEM professional development opportunities for surrounding districts.

Dr. Emily J. Klein is Professor at Montclair State University in the Department of Teaching and Learning and the Undergraduate Program Coordinator. She is also Academic Co-Editor of *The Educational Forum,* the journal of Kappa Delta Pi and co-Principal Investigator on the WIPRO Science Education Fellows grant that supports teacher leadership in New Jersey school districts. The author of several articles and books on teacher professional learning, teacher leadership, and urban teacher residencies, Dr. Klein's third, and most recent book, *Our Bodies Tell the Story: Using Feminist Research and Friendship to Reimagine Education and our Lives* came out in 2023.

Dr. Dorina Kosztin, Curators' Distinguished Teaching Professor, MU Physics & Astronomy, was a guiding member, co- Principal Investigator (PI) and PI of the Missouri Wipro SEF project from 2018 until her untimely death in 2023. It was her idea to focus on teaching science and math in a complementary manner for Wipro Phase 2. She began working on teacher professional development projects with the A TIME for Physics First project, which served about 120 teachers from 53 districts. She received more than a dozen awards for her teaching, including the prestigious Curators' Distinguished Teaching Professorship, and co-authored the Physics First curriculum and related Exploring Physics e-books. In the department she had major impacts on the introductory calculus-based physics courses, online teaching, and MU's curriculum committee, and beyond that, to teacher education in the state of Missouri.

Dr. Meghan Marrero is Professor of Secondary Science Education at Mercy University and co-Director of the Mercy Center for STEM Education, working primarily to improve access to STEM education for diverse learners. Her research centers on ocean literacy of students and teachers, as well as STEM teacher education. Meghan was a Fulbright Scholar to Ireland in 2018, where she focused on teaching and research around family learning in science and engineering for early childhood students and their families. Meghan holds a BS in Biological Science from Cornell University, an MA and Ed.D in science education from Teachers College, Columbia University, and an advanced certificate in educational leadership from Queens College.

Dr. Tammy Wu Moriarty is dedicated to enhancing the leadership capacity of teachers, school administrators, and district leaders, supporting their development by facilitating professional learning sessions and offering leadership coaching. Her background includes experience as a secondary science teacher, district science resource teacher, school administrator, and educational consultant. Dr. Moriarty holds a BS in Animal Physiology and Neuroscience from the University of California, San Diego, an MA in Educational Leadership, and a Ph.D. in Leadership Studies with a PK-12 Specialization, both from the University of San Diego.

Dr. Mika Munakata is Professor of Mathematics education at Montclair State University in the Department of Mathematics. She is the Principal Investigator for the Wipro SEF Fellowship program at the New Jersey site. Dr. Munakata is a former middle and high school mathematics teacher who currently teaches courses for preservice teachers, in-service mathematics teachers, and students in the Ph.D. program in Mathematics Education. Her research interests include interdisciplinary STEM education, undergraduate STEM education, and teacher professional development. She is especially committed to developing and implementing partnership programs with K-12 schools to support STEM teaching and learning.

Dr. Ratna Narayan, Associate Professor of Science Education is the Principal Investigator of the Wipro Science Education Fellowship grant at the University of North Texas, Dallas. A recipient of the Gerald Skoog cup for outstanding leadership in Science Teacher Education awarded by the Science Teachers Association of Texas Executive Board, Ratna enjoys teaching and mentoring pre-service and in-service teachers at the undergraduate and master's level. Her research interests include inquiry, STEM education, and informal science education.

Larry R. Plank is Assistant Professor of Science Education at the University of South Florida. He serves as the president of the Tampa Bay STEM Network, an original STEM Learning Ecosystem. Previously Mr. Plank served as the Executive Director for K-12 Science Education for Hillsborough County Public Schools, in Tampa, FL, where he directed the science education programs for over 220,000 students. Mr. Plank earned a Bachelor's degree in Biological Sciences from Florida State University in 1997, and Master's (Biological/Marine Sciences) and Specialist's degrees from the University of South Florida in 2000 and 2006, respectively.

Dr. David Rosengrant is Professor of STEM education and the campus dean at the University of South Florida. His research interests include physics education, virtual reality, eye-tracking technology, and video games as tools for teaching. He spearheaded the University of South Florida's St. Petersburg campus STEM INQ Lab, a state-of-the-art facility that features robotics, 3D printing, laser printing, coding, and virtual and augmented reality systems. The space allows aspiring educators and current faculty to infuse the latest in STEM into their teaching methods. He was named university-level Teacher of the Year in 2014 by the Georgia Science Teachers Association.

Dr. Marcelle A. Siegel is Professor of Science Education at the University of Missouri. She is jointly appointed in the Department of Learning, Teaching & Curriculum and the Department of Biochemistry. Her background in education includes several years in curriculum development for national reforms, professional development in urban and rural schools, and a focus on understanding and improving assessment practices in science classrooms. Continuously funded, Dr. Siegel currently serves as Principal Investigator of THRIVE an Inclusive Excellence initiative of the Howard Hughes Medical Institute and Co-PI of NSF ART's Mizzou TecHub.

Dr. Monica Taylor is Director of Gender, Sexuality and Women's Studies, professor in Educational Foundations, and doctoral faculty in the Teacher Education and Teacher Development Ph.D. program at Montclair State University. She received her Ph.D.

in Language, Reading, and Culture at the University of Arizona. She writes about feminist pedagogy, self-study, LGBTQ+ inclusive practices, teaching for social justice, and teacher leadership. Her newest book is *Our Bodies Tell the Story: Using Feminist Research and Friendship to Reimagine Education and Our Lives.* She is co-editor of *The Educational Forum,* and co-Principal Investigator of the WIPRO Science Education Fellowship grant.

List of terms

CCLS   Collaborative Coaching and Learning in Science; An approach to teacher development involving teams, videotaping of lessons, review of lessons by peers, and providing and receiving warm and cool feedback.

CoS   Course of Study; In the context of CCLS teams, the disciplinary content topic and pedagogical strategy or science and engineering practice to be investigated and implemented by the team.

CS   Corporate sponsor; The private entity providing the funding for the program.

CTS   Curriculum Topic Study; The disciplinary content idea to be explored by a CCLS team.

DSC   District Science Coordinator; Individual within a school district that coordinates the overall science education and curriculum strategy from pre-K through high school graduation.

GPS   Growth Plan System; An individual professional development plan for Fellows in Year 2 of the program intended to improve teacher leadership characteristics.

IHE   Institute(s) of Higher Education; Colleges and Universities.

IRB   Institutional/Investigational Review Board; A panel of personnel at an IHE tasked with ensuring ethical and legal compliance for all research sponsored by the IHE, particularly with respect to protection of human and animal research subjects.

NGSS   Next Generation Science Standards; A set of content standards, cross-cutting concepts, and practices developed to guide educators as they prepare curriculum for students in pre-K through grade 12.

NSF	National Science Foundation; A governmental agency that provides significant funding to research in the United States.
ORSP/OSP	Office of Research and Sponsored Programs/Office of Sponsored Programs; A group within an IHE that provides administrative (financial, contractual, legal, and clerical) support to research activities within the IHE.
PI	Primary Investigator; The individual at the IHE leading and ultimately responsible for a research project.
PLC	Professional Learning Community; A group of people, typically educators and/or administrators, working together for a common education purpose such as a book study group, a pedagogical study group, or project group.
PM	Project Manager; An individual tasked with maintaining and monitoring a project and its plan including operational goals and objectives, financials, and team performance.
SEF	Science Education Fellowship; A professional development and district transformation program to aid science teachers in developing as educators and teacher leaders.
SEP	Science and Engineering Practices; A group of skills and pedagogical approaches from the Next Generation Science Standards that students are intended to develop and master throughout their educational experience, pre-K through grade 12.

Introduction

Arthur Eisenkraft

This book, and more importantly the program it describes, focuses on developing science educators to their fullest potential. At its heart, the program is professional learning intended to bring about positive, sustainable district change through teacher leadership. The Wipro Science Education Fellowship (SEF) has existed since 2014. Hundreds of Fellows have completed the program and have impacted the lives of thousands of students in public schools, most of which are in high-needs school districts. It is a collaborative effort of teachers, science coordinators, school district administrators, and university professors and staff. In the pages that follow, the program will be described and explained, enhanced by testimonials of participants.

As educators, many of us have either been participants in professional learning or have been instrumental in the design of professional learning for teachers. What is it about this program, the SEF, that sets it apart and that teachers describe as a transformative experience? What makes the program so special that it has been successful in seven very different college and university sites and 35 school districts throughout the United States, with different higher education leadership, different school system dynamics, and different teacher backgrounds and expertise?

What we hope to do in this book is to share what we believe are aspects of the program that make it so special and at the same time provide a road map for educators that want to replicate the SEF program, in total or in parts, in their local area. Those educators may be university faculty that wish to support local districts' transformation through teacher leadership using the model of SEF. Alternatively, the educators may also be administrators, district coordinators, or teachers in the school districts who see elements of SEF that they can implement locally. Each university site and each school district we have worked with has had unique characteristics and found ways to implement the program for their locale. This book welcomes you to join us on this journey toward higher quality science education.

This first chapter of the book will provide the reader with an introduction to the program – what it is, what it hopes to accomplish, and how it is structured. This chapter will close with sage advice from Fellows that have completed the program.

DOI: 10.4324/9781003490586-1

Subsequent chapters of the book will provide more details on specific parts of the program. Chapter 2 will provide the reader with a summary and background of teacher leadership. This chapter intentionally will feel slightly more "academic" than the remainder of the book. Chapter 3 will dive into the first year of the program. The first year of the program is focused on improving instructional prowess. Chapter 4 will look at the second year of the program. This is the year of the program where Fellows will hone their personal leadership skills.

The second half of the book will look at different roles and responsibilities, first of the District Science Coordinator (DSC) in Chapter 5 and then the university faculty and staff in Chapter 6. A final chapter will describe the different paths that university sites and districts will navigate to meet the goal of district transformation. Since 2015, we have also engaged an independent evaluator (David Heil & Associates, Inc., from 2015–2020, and then Anne Gurnee Consulting, LLC from 2020–2024) to gather insights from program participants about the program's effectiveness and impact. Throughout the book, we will provide snapshots of key program evaluation findings that have helped to steer the program's development over the years and also to highlight the impacts on the program's stakeholders. Overall, the evaluation has continued to find that nearly all of the Fellow and the DSCs have been highly satisfied with the program, saying that it met or exceeded their expectations and made many positive impacts on their classrooms and districts.

We invite you to read the entire book or focus on the chapters that are most applicable. For example, if you are an educator looking to implement vertical (across grade bands, i.e. K-12) and horizontal (within a grade band, i.e. grades 6-8) curriculum articulations in your district, you may want to read Chapter 1 and then Chapter 3. The book is written as the program has been developed and deployed. The program itself is flexible to allow for variation to meet local needs.

The great Spanish poet reminded us: "Caminante, no hay camino, se hace camino al andar" (There is no road, we make the road by walking.) Many of us have been trying to create the road that will lead toward improved science education for all our students. We watch and learn from others who have been creating roads. Our hope is that our efforts, described in this book, will help guide you along your chosen path.

Chapter 1

The impact and structure of the Wipro Science Education Fellowship

Arthur Eisenkraft

The Wipro Science Education Fellowship (SEF) has profoundly impacted my teaching and leadership journey. I've had the honor of presenting at various levels, from campus to national platforms like National Science Teaching Association (NSTA), advocating for equal opportunity in education. These experiences, including serving on Texas Education Agency (TEA) committees and participating in Conference for the Advancement of Science Teaching (CAST), have given me a voice to champion diversity, equity, and inclusion for all students. They've inspired me to lead with empathy and determination, ensuring every student has the chance to achieve academically and thrive. My goal is to empower students to become productive members of the global community, particularly in the vital fields of STEM.

Candace Edmerson, High school, Grand Prairie, TX

The Wipro SEF program had a great influence on me as an educator, helping me to grow in confidence as a science educator and as a leader in my school. Since participating in the program, I helped my school to select a new science curriculum that we love and have led science instruction PD at my school. I feel much more comfortable collaborating with and leading other educators in my district and am now part of other committees to help improve our school.

Josie Hess, Upper elementary (3-5), Community R-VI Elementary, MO.

DOI: 10.4324/9781003490586-2

My overall experience with Wipro has been nothing short of amazing. When I first joined in 2018, I was a very shy person and not very outspoken in front of an audience. I always knew I wanted to help teachers but never had the courage to step out of my box. With the motivation of other Fellows, my DSCs, and Dr. Narayan, I am now doing what I always imagined while still transforming minds in the classroom.

Brittney Preston, Upper elementary (3-5), Lancaster ISD, TX.

These brief testimonials from teachers who have completed this program help to clarify the purpose of this book.

The Wipro SEF program is based upon the success of the Boston Science Partnership's Science Education Fellowship (SEF), which was supported through the National Science Foundation Math Science Partnership Program from 2009 to 2012. It is a district transformation program with the intention to recruit committed teachers of science[1] who are poised and ready to learn, to collaborate, and to be reflective on their classroom practice. The SEF program is designed to foster teaching and leadership skills in those who have a minimum of three years of teaching experience and are committed to staying in the classroom within their school district for at least the next two years – to lead without leaving the classroom. Fellows do not need to have prior leadership experience, and while many teachers may find leadership skills to be helpful to pursue other career avenues in the district, this program is not designed to facilitate the transition of a classroom teacher to becoming a school-based or district-based leader or administrator. Though some teachers do move their professional careers in this direction, as a result of experiences through the SEF, teachers who lead while remaining in the classroom are seen in SEF as success stories and our prime motivation and purpose.

To date, every instance of the SEF has been a partnership among an Institution of Higher Education (IHE) and local school districts. There are currently seven IHEs around the country (CA, FL, MA, MO, NJ, NY, and TX)[2] participating in the program. The IHE leadership team at each site comprises professors from Science, Technology, Engineering and Mathematics as well as Education. The program uses a model of teacher support and development to increase the quality of teaching and leadership in science throughout partner school districts. This model includes a comprehensive set of activities designed to improve teacher practice and effectiveness in the instruction of science. The IHE provides local leadership and resources to the program. The IHE also invites applications from and chooses the local partner school districts, with emphasis and preference given to high needs districts. This is certainly not to imply that a school district or group of school districts could not manage this program without a university partner.

Twenty teachers from grades K-12 are selected each year from the five partner school districts at each IHE site. These teachers (Fellows) have the opportunity to collaborate with district colleagues at other grade levels, their district science coordinator, IHE faculty and staff, and peers from other local districts. Each IHE site has a leadership team comprising science and science education faculty. At the conclusion of the four-year program, each of the five partner districts will have a cadre of 12 teacher-leaders who can support district initiatives and participate in leadership roles while remaining in the classroom.

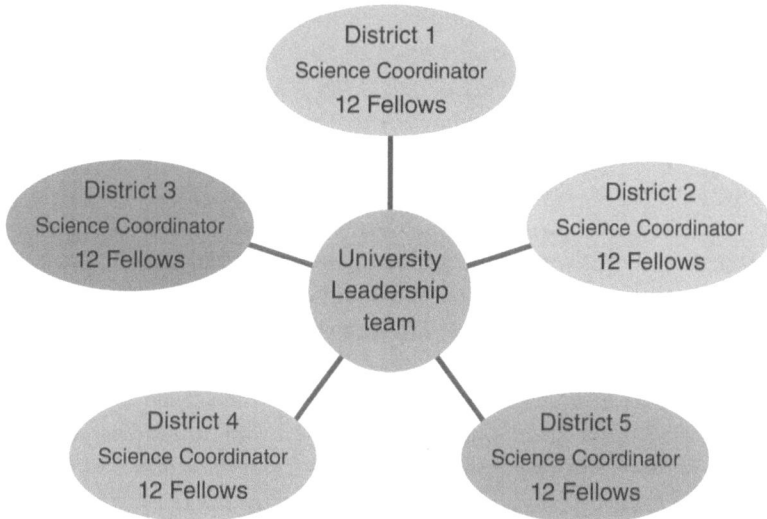

Figure 1.1 A pictorial representation of a local deployment of the Science Education Fellowship (SEF). The university leadership team and the science coordinators work together as the SEF leadership team.

Source: Image by the author.

Over the course of the two-year fellowship, district science coordinators and Fellows will collaborate not only with university leadership but with colleagues from the five partner districts. Cross-district collaboration is an important part of the program, providing Fellows and coordinators with perspectives they may not see in their "home" district. The program encourages districts to learn from each other as well as having universities learn from the school districts. The network of science coordinators provides a support system with colleagues who are facing similar challenges, and the university-school district partnership strengthens the relationship between university faculty, district leaders, and teachers (Figure 1.1).

Wipro SEF fundamentally changed the way I teach for the better. I didn't expect it going into the fellowship. It pushed me to uncomfortable places, sharing entire videos of imperfect lessons with my small groups and taking feedback. But it forced me to consider new ideas, observe new strategies in others' classrooms, learn about best practices, and have difficult conversations with other teachers who were also thinking deeply about their own teaching practice.

Kris Grymonpre, MIddle school, Boston, MA

The program is supported by three pillars:

- Reflective practice
- Adult learning
- Leadership

Each of these pillars manifests through intentional choices in the design of the program. If the end goal is for a Fellow to emerge as a teacher-leader, the starting point is excellent instruction. Literature has consistently shown that instructional leadership and being recognized as a strong teacher are critical for teacher leadership, particularly teachers without formal titles who lead from their classrooms. Being a reflective practitioner has been shown over and over to lead to improved instruction. The SEF program has built-in "stops" for reflection on this journey with routine journaling and reflection following instructional review protocols that will be described later.

The relationships among our V-CCLS team evolved and strengthened significantly throughout the semester. Initially, we came together with diverse backgrounds and perspectives, but as we collaborated on Biology 1 Claims, Evidence, and Reasoning, we forged strong bonds. Regular meetings and discussions provided opportunities to share ideas, experiences, and insights, fostering mutual respect and understanding. Over time, these interactions cultivated a supportive and collaborative environment where we felt comfortable exchanging feedback and supporting each other's growth as educators. Ultimately, the semester was characterized by camaraderie, trust, and a shared commitment to excellence in science education.

Candace Edmerson, High school, Grand Prairie, TX.

As a result of our H-CCLS semester, I realized that we all had different styles of teaching. I was open to trying different activities with my students and team. I realized that not all noise is chaos. I also learned additional locations for our students to learn outside of the classroom. The various locations can provide different experiences than those that arise in a traditional classroom setting. For example, students can develop critical thinking skills, problem-solving skills, and decision-making skills outside of the classroom.

La'Tanya Spragin, Middle school, TX.

The program recognizes that students will benefit greatly from their teachers' participation. The focus of the program is supporting and enabling the Fellows. As such, a focus on adult learning is critical. Drago-Severson and other scholars have provided a compelling case that adult learning is different enough from student learning so as to warrant separate emphasis and study. The Fellows begin by focusing on their own learning in the context of their instructional practice. They then begin to consider what they have learned about adult learning in the context of their colleagues when they begin to lead and deliver professional development sessions to their peers.

My experience with the program has taught me that adult learning environments, much like progressive student-centered classrooms, must be intentionally supportive to ensure universal access to learning for all participants. Specifically, when working with K-12 science teachers in the same space, it is crucial for all educators, especially elementary teachers, to feel equally valued within the group. The intentional structure should include frameworks designed to flatten sociocultural hierarchies that naturally exist.

Leana Peltier, High school, Tarrytown, NY.

It is then in this leadership of professional development where the third pillar emerges. Fellows will develop the skills necessary to lead initiatives and change efforts throughout their district. The program emphasizes distributed leadership, where teachers, science coordinators, and principals share strategies, goals, and visions for how the school and district should move forward. Fellows will study this theoretically through sessions led by the host university. They will also apply and practice these skills in small groups within their cohort and by leading real projects in their districts. Researchers have found that teacher learning as well as leadership is more about *peer influence* and is rarely "vested in one person who is high up in the hierarchy." (York-Barr and Duke, 2004). As Szczesiul and Huizenga (2015) have found, teacher leadership is a "socially distributed phenomena" that develops over time as teachers gain efficacy; to do so, they must have "repeated opportunities" to reflect on what they master in the context of structured collaboration. Whether in the form of well-designed lesson study or similar practices, this type of professional development helps teachers become more comfortable with "feelings of failure" and "cop[ing] with difficult situations."

Wipro connects me with teacher-leaders locally, nationally and even in other countries. This cross pollination of ideas has informed my practice and inspired me to get out of my comfort zone. The Wipro PLCs around School Gardens and Climate Change Education expanded my perspective and infused my ideas with relevance and nuance. Wipro helped our school complete a STEM garden that has energized our school community about outdoor learning. More than 30 teachers have collaborated on this project. For many of us, it was an opportunity to flex our leadership muscles beyond our classrooms.

Marcia Manzueta, Lower elementary (K-2), Portchester UFSD, NY.

Teachers from participating districts become a community of SEF Fellows through monthly meetings and Collaborative Coaching and Learning in Science (CCLS) cycles in Year 1 of the program. The university leadership team partners with district science coordinators to provide teachers with both the structures and the resources to work on pedagogical skills and research-based science education issues in a collaborative manner. Using classroom videos, the program allows the time and structure for Fellows to reflect on their own teaching and to provide feedback to their colleagues while building a supportive network of K-12 science educator colleagues. They carry this community into Year 2 by carrying out self-designed professional development plans (a "growth planning system" or GPS), while receiving guidance from advisors and their district science coordinators. In recognition of their work beyond the school day, Fellows receive either a stipend and/or tuition assistance over the two years as benchmarks of the program are met. This represents a commitment of 250 hours over two years (similar commitment to two or three graduate level courses). Each of the components is described in the remainder of this book. Table 1.1 describes the major milestones and interactions throughout the program. Table 1.2 highlights the major meetings and purpose of each.

There were some times that I dreaded the drive across town after a long school day, but honestly, I am glad that I made the effort. I left every meeting with a new idea or strategy, or better understanding of our work or project in progress. I understood that I was working

Table 1.1 Major events and milestones for SEF program

Year 1	Semester 1	Semester 2
Collaborative Coaching and Learning in Science (CCLS)	Vertical teams (organized by subject – life, chemistry, physics, environmental)	Horizontal teams (organized by grade level – elementary, middle, high school)
CCLS presentations	End of 1st semester	End of 2nd semester
Monthly university professional development	10 meetings	10 meetings
End of year conference		Fellows from all districts

Year 2		
Independent professional development plan ("growth planning system" or GPS)	Establish district and personal goals for GPS and begin projects	Complete GPS including participation with other teachers in the school
Professional development to peers		Professional development on findings from GPS are presented to colleagues, university, and school district partners.
GPS poster		
GPS portfolio	Artifacts and reflections on GPS	Final GPS portfolio

with a select group of educators, and every one of my team members came prepared and ready to contribute. I was motivated to bring my best effort each time we met. Some of the most memorable meetings included the larger science community, representatives of the zoo, arboretum, and science museums. I met so many people from across my district and other districts, and made so many valuable connections with the larger community of science educators. I continue to utilize what came out of those meetings to enhance my classroom instruction.

James Mining, Middle school, Irving ISD, TX.

Distributed leadership

Distributed leadership has become popular within education over the past 15 to 20 years. Distributed leadership looks at leadership from the perspective of the process of leadership and through the interactions of the people involved in leading. It is not a theory that emphasizes characteristics of individuals, but rather emphasizes the practice of leadership. This will be discussed further in Chapter 2.

From a district perspective, the first cohort comprises four Fellows from that district. These Fellows work with the other 16 Fellows of Cohort 1 (four each from the other four local districts, for a total of five districts and 20 Fellows). When Cohort 1 begins their GPS in Year 2, the second Cohort of four Fellows begins Year 1 of the program.

A Roadmap for Transformative Science Teacher Leadership

Table 1.2 Key meetings and goals for meetings for SEF program

Type of meeting	Who leads?	Focus of discussion and activities/pillar	Who attends?	Frequency
Monthly cohort meetings	IHE team with input from District Coordinators	Reflective practice/Adult Learning/Leadership	Year 1 Fellows, District Science Coordinators, SEF staff (site specific)	10 times (8 after school, 2 Saturday)
Vertical Collaborative Coaching and Learning Science (V-CCLS) meetings	Fellows take turns leading	Reflective practice/Adult Learning/Leadership	SEF Fellows by Subject Vertical (V-CCLS) teams	5–7 times in the Fall of Year 1 Scheduled by vertical team members
Horizontal Collaborative Coaching and Learning Science (H-CCLS) meetings	Fellows take turns leading	Reflective practice/Adult Learning/Leadership	SEF Fellows by grade span (H-CCLS) teams	5–7 times in the spring of Year 1 scheduled by horizontal team members
District cohort meetings	District Science Coordinators	Adult Learning/Leadership	District Science Coordinator and their Fellows	3–4 times per year, in both years of cohort fellowship and afterward (voluntary)
Advisory meetings	IHE Advisors	Adult Learning/Leadership	Advisors and their assigned Fellows	6 times during Year 2 of cohort fellowship
District coordinator Professional Learning Community (PLC) meetings	District Science Coordinators	Adult Learning/Leadership	District Science Coordinators only	4 times per year*
Leadership team meetings (one site)	IHE leaders and District Science Coordinators	Reflective practice/Leadership	District Science Coordinators and SEF IHE staff	5 times per year* (try to coordinate with monthly cohort meetings)
Cross site meetings and conferences				
Leadership cross-team meetings (multiple sites from across the country)	IHE lead institution	Reflective practice/Leadership	District Science Coordinators and SEF IHE staff from all sites	1 time at the end of Year 1 for each cohort
Cross-site SEF teacher leadership conference	IHE lead institution	Reflective practice/Adult Learning/Leadership	All Year 1 Fellows, District Science Coordinators, SEF IHE staff from all university sites	1 time at the end of Year 1 for each cohort
Cross-site SEF leadership meeting	IHE lead institution	Monthly updates and sharing across sites	IHE from participating universities	Monthly

* Frequency of leadership team meetings and coordinator PLC meetings can change after program is fully established.

Table 1.3 The SEF program evolves over four years to ensure that each district will have a critical mass of Fellows

District	A	B	C	D	E	Total for All Districts
Cohort 1	4	4	4	4	4	20
Cohort 2	4	4	4	4	4	20
Cohort 3	4	4	4	4	4	20
Total for all districts	12	12	12	12	12	60

Note: Each cohort participates for two years. The number of Fellows and the number of districts may vary for very large or very small districts.

The following year, Cohort 2 Fellows work on their GPS and Cohort 3 Fellows begin Year 1 of the program. Cohort 1 Fellows have formally completed the two years of the SEF program, but are expected in subsequent years to interact with Cohort 2 and Cohort 3 Fellows. At the conclusion of the four years, each district has 12 Fellows across elementary, middle, and high school who have completed the two-year program and the SEF requirements, as shown in Table 1.3.

They have all learned a common approach for improving instruction, reflecting on research, understanding how curriculum can be vertically articulated and how science and engineering practices can be integrated into their science content. They have also each participated in a GPS, part of which has focused on working with other teachers in the district and supporting district initiatives.

At the conclusion of the SEF program, the District Science Coordinator (DSC) is no longer alone in the work of leading science within the district. They now have others who know them and understand the work of the DSC – and can help carry the work forward. The DSC now has 12 Fellows that can support and critique future district science initiatives. This new distributed leadership across the district provides the opportunity for sustainable improvements in K-12 science education in the participating districts long after the SEF program has concluded. The process of distributing leadership becomes second nature to the Fellows. In fact, at horizontal Collaborative Coaching and Learning Science (H-CCLS) and vertical Collaborative Coaching and Learning Science V-CCLS presentations, when asked "who was the leader of your group," all of the Fellows will look at you with confusion. They recognize that they weave in and out as the leader of the group. By design, during the CCLS cycles, teachers develop their leadership skills through practice as they rotate through roles and share responsibilities. Additionally, there are frequently opportunities for distributed leadership in their GPS work as they bring together others from their districts in the professional learning that they have developed.

As a Fellow in Wipro SEF, I had the opportunity to engage in district work as a teacher. At that time I was asked to evaluate and provide feedback on some of the lessons being used to teach science in the Kindergarten classrooms. The Science and Early Childhood departments listened to my feedback and actually took it into consideration. This had a big role in my move into district leadership.

Now, as a district leader, we strive to create ample opportunities for our teachers to grow their leadership. We have created a "Science Stewards" program for our champion science teachers. Our Science Stewards work with the STE department throughout the year to lead

A Roadmap for Transformative Science Teacher Leadership

professional learning, provide feedback on district work, and help us build a community throughout Science PLCs.

Molly Peters, Lower elementary (K-2), Boston, MA.

In my current school district, they created committees to allow teachers and district leaders to work together to make informed decisions. As a teacher, I was asked to become a science lead teacher and to join several committees to make decisions about the campus and student engagement. I was also able to work with my district science coordinator on science curriculum and professional development for the school district.

Raisha Allen, Middle school, Desoto ISD, TX.

Key components of the SEF program

Corporate sponsorship

The SEF program has been funded since August 2012 by a global information technology services[3] corporation. The company has committed more than $18 million to the Center of Science and Math in Context (COSMIC) at the University of Massachusetts Boston as the lead university to partner with other universities in the United States to implement the program for three cohorts of 20 Fellows at each of the seven sites for a total of over 500 school teachers in districts near each university site: the University of Massachusetts Boston (MA), Montclair State University (NJ), Mercy College (NY) the University of North Texas Dallas (TX), Stanford University (CA), University of South Florida (FL), and University of Missouri (MO). Much of the money went to teacher stipends. The current allocation of funds has been targeted to district transformation innovations after the core SEF program of four years was implemented at each site.

This model of partnership, where STEM scientists and educators (IHE team) at a college or university partner with a corporate sponsor in local high-needs school districts, can be replicated across the United States using the SEF model. Without minimizing the impact of corporate support, we also think that there are ways in which the program can run without corporate sponsorship and well within the professional development budgets of local districts. In these situations, districts use alternative incentives to promote participation.

University and public school district partnership

Each university site maintains the program and engages each of their partner school districts in the work of the SEF. Generally, there will be five partner school districts to facilitate a large enough pool of Fellows with unique perspectives and experience. In very large districts with many schools, fewer districts can provide this. Conversely in smaller, usually rural area, more than five districts may be necessary to get a large enough applicant pool. Through the partner district application process, each of the partner districts selected must make a commitment to have their DSC actively work with the university to implement the program. The DSCs are important because they are a liaison between the Fellows and the SEF university staff: they help to recruit and to select teachers for the program, they attend monthly meetings, and they assist in the planning of the program. They provide invaluable insight into district priorities and initiatives in which the SEF program is able to complement and support. They also provide important professional development and leadership to their teacher Fellows. Finally, it is the DSC who will find ways to provide opportunities in future years for the Fellows to contribute to sustained district success.

Working across partner districts

One of the strengths of the SEF program is the fact that the Fellows, the DSCs from the five partner districts, and the university personnel become a Professional Learning Community (PLC) within the SEF program. The SEF Fellows and DSCs from the partner districts find that they have similar needs and common goals. Fellows, through the V-CCLS and H-CCLS communities state they feel less isolated and are able to accomplish more as reflective practitioners through the sharing of classroom videos and by providing feedback and teaching insights. All participants, including the university staff, gain more familiarity and build relationships with the neighboring partner districts.

Supporting and incentivizing the work of teacher-leaders

The Science Education Fellows grow to be teacher-leaders by participating in a program that focuses on the teacher as a reflective practitioner, an adult learner, and as someone who is able to learn to lead from the classroom. They can serve as leaders while they remain committed to teaching in the classroom. Teachers become a community of SEF Fellows through monthly meetings and CCLS cycles in Year 1 of the program, and then by carrying out self-designed professional development plans (GPS), while receiving guidance from advisors and their science coordinators in Year 2 of the program. The program allows the time and structures for the Fellows to reflect on their own teaching and to provide feedback to their colleagues while building a supportive network of K-12 science educator colleagues. In recognition of their work beyond the school day, Fellows receive a stipend over the two years as benchmarks of the program are met. The stipend is monetary compensation for a commitment of 250 hours (outside of the school day). As the present participating school districts and IHEs are now receiving diminished external funding, they are now exploring alternatives to stipends (e.g., release from some duties, course release, salary lane credit, in-service or IHE course credit, district/school/release time, union contract sanctioned professional development time, etc.)

While monetary incentives are great, research has shown that they do not necessarily lead to sustained change in teacher leadership. Tapping into Fellows' intrinsic motivation will likely have more lasting results.

Collaborative Coaching and Learning in Science communities (CCLS)

The CCLS model was born out of the Boston Science Partnership program (an NSF math-science partnership) and based in the practice of "Observation Cycles," which uses videos of classroom lessons for the purpose of changing practice through honest and open reflection.[4] In SEF, Fellows participate in both V-CCLS and H-CCLS communities in Year 1 of the program. The community creates a CCLS culture by setting group norms so the sharing of classroom videos is done in a safe and supportive environment and keeps the focus on reflecting upon the team developed Course of Study (COS) including science content, science and engineering principles, and research articles. During the CCLS cycle, the CCLS communities develop a common language, engage in respectful conversations, and follow specific protocols when viewing and providing feedback to each other's classroom videos. More about the V-CCLS and H-CCLS communities and timeline is detailed in Chapter 3.

Sharing Collaborative Coaching and Learning in Science research

Fellows are given two opportunities to share what they have learned in their CCLS communities and through their classroom research and reflection in Year 1. In January, at each university site, the Fellows present their V-CCLS findings to their peers through 20-minute presentations and then a round of warm and cool feedback. This could be considered "practice" for the H-CCLS presentations, which are given at the regional SEF Teacher Leadership Regional (cross-site) Conference in June of Year 1. The regional conference provides an opportunity for the teachers from different university sites to network and build relationships with others who have gone through a similar but not identical program. The SEF Teacher Leadership Regional Conference is a professional learning environment that is supportive, diverse, and collaborative for all involved: the Fellows, the DSCs, and the university faculty and staff.

Individual Growth Plan Systems (GPS)

At the end of Year 1, Fellows are afforded the rare opportunity to design an individual GPS, allowing them to create their own learning pathways, both personally and professionally. The plan consists of a clear vision of how the Fellow can support district initiatives in his/her school and district while also defining personally important work tied to improving their own instructional practices. At least 30% of their Fellowship time in Year 2 is devoted their personal growth and at least 30% of their Fellowship time is to be focused on professional learning and in leadership activities supporting district initiatives. The remaining 40% of their time is distributed between these two areas (personal and professional growth) as the Fellows see fit to best support their plan.

The GPS is a well-defined plan that shows where each Fellow would like to be at the end of Year 2. Fellows collaborate with an IHE Leadership GPS Adviser and their DSC to develop their own professional growth plan, which is supported through Year 2 by them and the other Fellows. Fellows discuss how to lead from the classroom while supporting a district initiative. The GPS elements and GPS projects are detailed in Chapter 4.

Conclusion of the chapter/beginning of your journey

In the preceding pages, we have described the major elements of the SEF. We have shared what we believe is what makes the program special and different from other professional development programs. The remainder of the book will provide more details on each aspect of the program and highlight work of previous Fellows. The book will help universities on their journey to support science education in their schools; it will help school districts who embark on this journey with or without the university leadership; it will help district science coordinators and teacher-leaders who will travel together in their schools. As we close this first chapter, you may be wondering if all of this work is "worth it" – does it really engage teachers and improve their practice? We have included participants' "words of wisdom" throughout the book, to answer these and other questions. At the completion of the program we asked Fellows, "What advice would you give to incoming Fellows?" and they shared their responses in the form of "Dear Incoming Fellow" letters. We invite you now to read and reflect on a few responses as you consider the program further.

Dear Incoming SEF Fellow......

To a teacher starting the Wipro SEF program, I'd suggest embracing the collaborative nature of the experience. Engage actively with fellow educators, sharing insights and learning from their perspectives. Be open to feedback, as it's essential for growth. Take advantage of program resources and support, including mentorship opportunities and access to research-based practices. Reflect regularly on your teaching and leadership goals, and be proactive in seeking out opportunities for professional development. Above all, approach the program with curiosity and a commitment to lifelong learning, and you'll make the most of this transformative opportunity.

Candace Edmerson, High school, Grand Prairie ISD, TX.

My advice to a new Wipro Fellow is to embrace this community of professionals and all the experiences and support it provides. Take the risks you have always been hesitant to take. You will not fail. You will grow professionally by leaps and bounds by forming new social networks and being inspired by other passionate educators. The Wipro Science Education Fellowship program opens a vast number of opportunities. Take them and enjoy every moment!

Susan Bartol, Upper elementary (3-5), Montclair, NJ.

From KC, Grade 6 Science/Inclusion Teacher

Congratulations on winning this fun and rewarding learning adventure! So many good things will come from this experience ... a better understanding of yourself as a teacher; a bunch of new research avenues to explore; many laughs with new friends; a new understanding of how the art of science teaching starts, evolves, and finishes from your colleagues; and a new confidence about yourself as you grow and grow and grow some more in your scientific expertise!

Good luck, have fun, and be prepared to compromise and share and listen!! Oh, and don't miss the NSTA conference wherever it may be....it's wonderful!

From KS, Grade 5 Science Teacher

As UMASS/SEF Fellows, we have a unique opportunity to collaborate with teachers, administrators, and professors who all share the same goals and visions as science educators. We are able to spend quality time planning, discussing, and presenting the strategies and practices which are not only going to teach our students science but to train them to think like scientists.

From KB, Grade 9 Earth Science Teacher

Things I didn't expect to happen, but did:

1   Although it was great hearing the "warm" feedback about my classroom, I was way more excited about the "cool" since it gave me so many ideas.
2   I wasn't as nervous about videotaping myself, knowing others would watch it, as I thought I'd be.

From EF, Grade 10 Chemistry Teacher

Don't be afraid to try something new and take risks, you can sometimes get much more out of debriefing a lesson that doesn't go as perfectly as you would have liked than one that did.

SEF Advice from Cohort 1 to Cohort 2: It's all about your partners and the team you build.

From GP, Grade 9–12 Physics Teacher

The relationships you build with your cohort partners will determine whether you enjoy the experience or not. Strong partnerships lead to strong leadership teams. In your cohort are your future collaborators. So here are five tips for building happy relationships with them so you'll want to continue working together in the years ahead.

1   Pick your dates early and stick with them. Everyone in your teams will be busy like you with calendars full of home obligations, out-of-work professional opportunities, and schoolwork. Changing dates after they've been agreed upon just causes tension. If something comes up and you absolutely need to change, communicate with your partners early, be friendly, and bring them chocolate.
2   Get your assignments done on time. It's unfortunate and unfair when a partner doesn't do their part. Everyone is making sacrifices to be in the program and everyone deserves to get all of the feedback and fresh ideas they signed up for.
3   Bring food. Food always helps to grease the partnership wheels. It can be as simple as a bag of trail mix to share, or as elaborate as an organized potluck. Food tells your partners that you care. While we're at it, drink doesn't hurt either.
4   Take your partners seriously. You, and everyone in your cohort, have been selected because you are serious about your work as an educator. You all want to learn, grow, and share, and you are all good at what you do. Take advantage of the opportunity to learn from and with them.

5   *Don't take yourself too seriously. Enjoy the experience and have fun with it. You will stretch and grow easier if you are flexible, and more people will want to partner with you. Good luck!*

RC Grades 6--8, Science Teacher

Your SEF experience may feel awkward and intimidating at first. You're opening up your classroom to people you don't know and asking them to tell you what they see. Scary, huh? But, trust the experience. You'll be working with a cohort of educators who are there because, like you, they want to improve their craft in a challenging and safe setting, with the goal of becoming better teachers for their students. The more open and honest you are, the more you will get from SEF, and the more you will grow as a professional. Enjoy!

Notes

1   Most K-5 teachers are customarily not "science" teachers but are generalists. They are an important component of the SEF program. For purposes of vertical curriculum articulation teaming, K-5 teachers are placed into a science subject area based on several factors that aim to create balanced vertical teams. When placing teachers in vertical teams, consideration is given to the distribution of teachers across districts and issues of equity regarding gender and race.
2   Stanford University (CA), University of South Florida, University of Massachusetts Boston (MA), University of Missouri, Montclair State University (NJ), Mercy University (NY), and University of North Texas, Dallas.
3   Wipro Limited (NYSE: WIT, BSE: 507685, NSE: WIPRO) is a leading technology services and consulting company focused on building innovative solutions that address clients' most complex digital transformation needs. With more than 230,000 employees and business partners across 65 countries, they deliver on the promise of helping their customers, colleagues, and communities thrive in an ever-changing world. For additional information, visit www.wipro.com.
4   This, in turn, was inspired by the Boston Plan for Excellence (BPE) https://www.bpe.org/bpe/about-bpe/our-history/

References

Szczesiul, S. A., & Huizenga, J. L. (2015). Bridging structure and agency: Exploring the riddle of teacher leadership in teacher collaboration. *Journal of School Leadership*, *25*(2), 404.
York-Barr, J., & Duke, K. (2004). What do we know about teacher leadership? Findings from two decades of scholarship. *Review of Educational Research*, *74*(3), 255–316.

Chapter 2

Vision and implementation of district transformation through teacher leadership

Arthur Eisenkraft, Amanda Gunning, and Meghan Marrero

The vision of the Wipro Science Education Fellowship (SEF) program is district transformation *from* the classroom – literally grassroots-type of work where teachers use their intimate knowledge of the learning context and students to enact meaningful change in and beyond their classrooms. Whereas some posit that the idea of district transformation is grandiose or too difficult, this kind of teacher-led approach is both doable and effective. The SEF program's vision is district transformation through teacher leadership. District transformation is a compelling rallying cry that can intimidate some people who interpret this as a grand DISTRICT TRANSFORMATION. Those familiar with school district change know that no single initiative can affect this level of change across an entire district. In this context, district transformation is presented as a set of smaller initiatives that nudge the district to changes in culture, direction, and goals. These incremental changes, led by the people who are in the classrooms each day, can guide a school district down a new path.

Regardless of how we achieve district transformation, we can be certain that it comes about through the efforts of teachers, administration, students and the community. SEF's focus is on developing teacher leaders – teachers who can lead without leaving the classroom (Whitworth et al., 2022). The teachers who

DOI: 10.4324/9781003490586-3

participate in SEF want to take on leadership roles but do not want to leave the joy of the classroom. Too often, the only path available for leadership roles requires a teacher to take on administrative responsibilities and titles (Natale, C.F., et al., 2013). We have found that this does not have to be the case. Teachers have stepped up and worked on projects that excite them and simultaneously support district initiatives.

There are a host of teacher leadership programs (e.g., Knowles Teacher Initiative, The New York State Master Teacher Program, Robert Noyce Master Teacher Fellowship) that aim to identify, support, and encourage individual teacher leadership. In programs such as these, teachers develop leadership skills and choose to become leaders in their schools, in their states, or nationally. They develop curricula and lead workshops as well as present at conferences.

The SEF program is unique in that it creates a cadre of teacher leaders within a particular school district. These K-12 teachers form a critical mass and network of teacher leaders who can support district initiatives. Our independent evaluations have regularly reported that teachers have increased their confidence and their self-concept as leaders. They have also improved leadership skills and communication behaviors like giving and receiving feedback. They also regularly report becoming more reflective practitioners who turn more regularly to research to guide their practice. They have worked with each other, their district science coordinators, and university professors and staff. They have participated in a common program and speak the same language. They know how to critique each other's lessons and to provide meaningful feedback.

They have experience with organizing and facilitating professorial learning experiences for colleagues and others beyond (outside of) the SEF program.

In addition, this cadre of teacher leaders from a single district have worked during their two-year fellowship with a cohort of other teachers from four neighboring districts. These interactions are a means to recognize what features of science teaching are unique to single districts and which are common across districts. Sometimes it is easier to support and learn from a teacher who is from a neighboring district and/or from a different grade level. Information and techniques for dealing with challenges or garnering other kinds of support can be shared from district to district. Commonalities are found within classrooms that are across town or in different grades. The teachers find that they share so much that it becomes easier to share more.

We keep searching for the "special sauce" that creates the success of the SEF program for the Fellows, teachers, district science coordinators, and school districts. Is it that the program supports teacher leadership of K-12 teachers, and the elementary, middle, and high school teachers learn from one another? Is it that neighboring school districts work side by side? Is it specific features of the two-year program? Is it that we have a group of teachers striving to be leaders who work with each other? Is it a relationship that bonds the teachers with each other and/or the teachers with their district science coordinators? Is it that taking the risk of opening their classrooms to strangers and being vulnerable forges relationships that last? Is it the respect given to teachers to create and implement a Growth Plan System (GPS) that includes both personal and professional goals? Could it be the content-specific professional community that few teachers experience in their schools and districts? These questions form the basis of our continued research on SEF.

I have become more confident as a professional among my peers. I have also gained confidence and experience in co-creating a mutually beneficial project with my peers, which I believe is one of the best ways for a teacher leader to affect their district. (MA teacher)

I think that the combination of the vertical and horizontal groups helped me to get a better perspective of how science is presented in other grades. I hope this knowledge will give me a bit more credibility with my fellow teachers of all grade levels as I try to take more of a leadership role in my own district. I also think that developing my skills in the area of public speaking will be a help. (NJ teacher)

As many of us discussed at the beginning of the program, our peers may take exception to one of their own stepping out into a leadership role. I am working towards that goal by speaking to fellow teachers in small group settings where we can all be part of the conversation, but I can still share what I have been learning through [SEF] and how important I believe it is to be prepared for the future of education. It has been received fairly well so far. (NJ teacher)

Teacher leadership is well-documented to promote positive educational changes for students (National Research Council [NRC], 2012). Understanding how to develop teacher leaders is a potentially powerful means to scale quality STEM education. The SEF model focuses on distributed leadership at the district-level and emphasizes (1) changes in teachers' instructional practices; (2) the development of teacher leadership; and (3) the spread of distributed leadership across the school district during and after participation in the program. A key is that the SEF model does not seek to change teachers but rather supports them as they transform their practices – individually and collectively. As of now, more than 500 teachers from 35 school districts in seven regions of the country have helped us identify salient components of leadership within schools and districts that are transferable to other settings. Major elements of the program model also have been successfully implemented in parallel settings with great success, showing this has promise for supporting change.

Science teachers need access to high-quality, sustained opportunities for professional learning. Unfortunately, most teachers engage in very little professional development (PD) specific to science (Banilower, 2019). For science teachers to update their knowledge and strengthen their practice in support of their students' learning, they will need professional learning opportunities that are sustained and collaborative (National Academies of Sciences, Engineering, and Medicine [NASEM], 2015). The SEF presents an answer to that call. It provides opportunities for teachers to work collaboratively to advance their understanding of the *Framework for K-12 Science Education* (NRC, 2012) and *Next Generation Science Standards (NGSS)* (NASEM, 2015). A team of local leaders who understand the necessary changes, are aware of district priorities, and have the skills to work with other teachers is essential for school reform (Gess-Newsome et al., 2009).

The remainder of this chapter is presented against the educational backdrop just described. Compared to the other chapters of the book, this chapter will read a little more formally and "academic." The intention is to provide a common background on what work has been undertaken in the area of teacher leadership by scholars and practitioners. This chapter also situates the SEF program in the larger context of teacher leadership scholarship and highlights contributions that the program can share with the greater education community. We will first offer our proposed Theory of Action, which strives to describe how teachers develop as teacher leaders. We then support this with related research from the larger field of teacher leadership including a discussion of distributed leadership. We will share some examples of how teacher leadership manifests in schools.

Teacher Leadership can be defined as "the process by which teachers, individually or collectively, influence their colleagues, principals, and other members of the school community to improve teaching and learning practices with the aim of increased student learning and achievement." (York-Barr & Duke, 2004)

Teacher leadership encompasses the practices through which teachers – individually or collectively – influence colleagues, principals, policy makers, and other potential stakeholders to improve teaching and learning. (Holland et al., 2014)

We believe teachers are leaders when they function in professional learning communities to affect student learning; contribute to school improvement; inspire excellence in practice; and empower stakeholders to participate in educational improvement. (Childs-Bowen et al., 2000)

In those instances, within the literature in which researchers did define teacher leadership for their study, it was conceptualized as working beyond the classroom walls, supporting professional learning in their schools, and being involved in policy-making and decision making at some level with the ultimate goal of improving student learning and success and seeking improvement and change for the whole school organization. (Wenner & Campbell, 2017)

Theory of action

Whitworth and Chiu (2015) proposed a model of how leadership may influence teacher learning and practices, and ultimately student achievement as illustrated in Figure 2.1 (solid arrows only). This model emphasizes the critical role of district and school leaders in supporting the planning and implementation of PD. The SEF model adds additional pathways to this figure (the dashed arrows). In the SEF program, we posit that leadership must be distributed throughout a school district to sustain efforts that positively influence teacher practice and therefore impact the learning of students.

During the SEF program, teachers (who are called Fellows) participate in PD programming, improve their knowledge and skills, and then change their practice

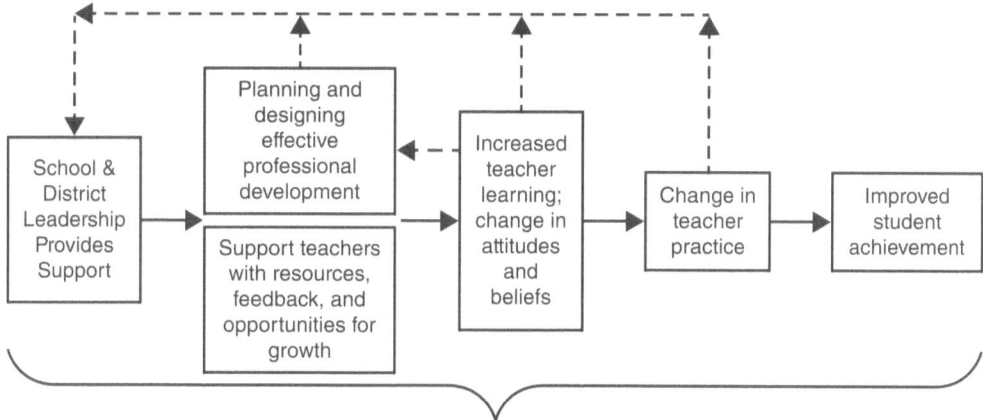

Context such as teacher, and student characteristics, curriculum, policy and working environment

Figure 2.1 SEF Theory of Action – Teacher leaders can leverage their knowledge and skills and become part of district leadership to plan and lead effective professional development to support their fellow teachers with resources, coaching, and feedback, thereby becoming a distributed leadership model.

Sources: Image by Whitworth and Chiu (2015) and modified by the author.

(Desimone, 2009; Luft & Hewson, 2014). The Fellows also gain leadership knowledge and skills that contribute to their school and district vision and that support their fellow teachers with professional learning opportunities, coaching, and feedback. They then become the future of their schools and districts as leadership contributors and providers.

The SEF program supports teacher learning by creating collaborative communities that support the learning of teachers and, ultimately, students. As teachers work collectively, they refine their knowledge and abilities for teaching and leadership while expanding their reach and distribute their leadership abilities. In this distributed leadership model, we view the school and the district as the units of analysis, with leadership distributed among teachers and supported by the administrators (Spillane et al., 2001). A key construct of this form of leadership is to view the social context of school as the conduit for activity that enhances both teacher and student learning (Spillane et al., 2001).

An important component of distributed leadership is the decentralization of the leader. In this approach, there is no one leader who is solely responsible for guiding the school. Instead, the leadership is fluid – drawing upon whoever is in the best position to offer guidance (Harris, 2003). According to Harris (2003), this orientation is important for three reasons: it allows multiple individuals to work collectively to guide instructional change, it allows for the work of leadership to be spread over multiple leaders, and it promotes a culture of shared responsibility rather than dependency. Across all of these areas, school personnel work together while building individual agency, which results in student learning.

Our theory of action is that by focusing on teachers who are willing to examine their instructional practice and take on leadership practices, they will develop into teacher leaders with high self-efficacy. In the SEF program, teachers participate in professional learning experiences that deepen their understanding of teaching and learning and support colleagues choosing to change their practice (Desimone, 2009; Luft & Hewson, 2014). The SEF Fellows gain leadership knowledge and skills and then plan and conduct effective PD and support their fellow teachers with resources, coaching, and feedback. With approximately 12 teachers working across a district with a common purpose, distributed leadership becomes a mechanism for sustainable support.

Related research

Science teacher leadership

Cultivating science teacher leaders is important. However, the reality is that science teachers are not typically engaged in opportunities to improve their pedagogical practice and leadership skills. For example, it has been documented that science teachers do not have ample opportunities to engage in sustained PD programs. Banilower (2019) found that 65% of teachers spent 35 hours or less over a three-year period engaged in PD programming related to science teaching. The SEF program engages teachers in more than *three times* this amount of professional learning during *each year* of the two-year program. When science PD programs are offered, science teachers often report not having access to specific types of programs such as opportunities for teachers to build their leadership (Luft et al., 2009).

Corresponding to the limited opportunities to build one's leadership capacity are the limited studies about science teacher leaders (Whitworth et al., 2022). Most of the studies are focused on a small group of teachers and describe a specific aspect of being a leader rather than a holistic look at developing leadership in teachers. Hofstein, Carmeli, and Shore (2004), for instance, focused on how teachers learned to change the teaching of science in their schools. Howe and Stubbs (2003) examined how a small group of science teachers grew as leaders during a PD program.

Preparation of science teacher leaders has increased in recent years (e.g., Whitworth et al., 2022). Some programs engage smaller groups of science teacher leaders (15–30 teachers) over the course of varying years of professional learning (Lotter et al., 2020; Yow et al., 2021). One program spanned seven years of PD for science teacher leaders and included graduate coursework, problem-based learning training, support to create school-wide STEAM nights at their home schools, and presenting at state science teacher conferences (Barth et al., 2023).

Similar professional learning to train science teachers to become leaders in their school and district include graduate coursework for participants to gain a teacher leadership endorsement from the state board of education (Criswell et al., 2018). Along with gaining leadership skills, participants engaged in professional learning communities (PLCs), mentoring activities, and furthering their pedagogical and content knowledge. While both of these programs engage a smaller number of science teachers, the aim for each program is for the participants to return to their home schools and engage their colleagues in further building teacher leaders.

More recent work by Sinha and Hanuscin (2017) and Hutchinson and colleagues (2023) explored the development of science teacher leaders and how they develop their identities as teacher leaders. Sinha and Hanuscin (2017) found that teacher leadership was a result of the priorities of the teacher, the school context, and the teacher's life experiences. This view of leadership resulted in teachers expressing their leadership abilities in a variety of different and diffused ways, and suggests many teachers can be leaders. Participants in Hutchinson and colleagues' (2023) study experienced tensions across the groups they interacted with as teacher leaders. Teacher leaders reported unsupportive principals and undefined leadership responsibilities as major roadblocks in supporting colleagues and student learning. The community gained through the PD enabled teachers to push through isolation of leadership and gain support from others in similar positions.

Outside of these studies, reviews of research on general teacher leadership offer insights in teacher leaders' development. York-Barr and Duke (2004) looked at two decades of teacher leadership. While they described different aspects of leadership, they also suggested how teacher leaders emerged in school settings. They specifically concluded that all teachers develop their leadership abilities and influence their schools and districts in different ways. Some teachers excelled at instructional leadership, while other teachers excelled at mentoring or coaching. Similar findings were identified in the decade following York-Barr and Duke's work by Wenner and Campbell (2017). Galosy et al. (2017) concluded that most of the findings of general teacher leadership were applicable to the science teacher community.

In the interim, a group of educators began to meet and discuss teacher leadership with their work culminating in 2011 with the release of the Teacher Leader Model Standards by the Teacher Leadership Exploratory Consortium (Cosenza, 2015). These standards define seven domains of teacher leadership (Teacher Leader Model Standards, 2011). The domains are listed in Table 2.1 (Teacher Leader Model Standards, 2011, p. 9).

One can see that the domains cover a wide spectrum of potential teacher leadership activity. The report and standards reinforce the ideas that teacher leadership is a term with diverse meaning that can be both informal or formal, and from the classroom or outside of a classroom (Teacher Leader Model Standards, 2011). Table 2.2 provides short descriptive examples of how the SEF maps to these seven domains.

Across the studies of general and science-specific teacher leadership, there is general agreement that more research in-depth, over time, and across larger populations is needed (Galosy et al., 2017; Wenner & Campbell, 2017; Whitworth et al., 2022; York-Barr & Duke, 2004). Wenner and Campbell (2017) note specifically that questions focusing on how teacher leadership is enacted and how school-level factors influence that enactment should be included in future work.

Table 2.1 The seven domains of the Teacher Leader Model Standards

Domain	Description
I	Fostering a collaborative culture to support educator development and student learning
II	Accessing and using research to improve practice and student learning
III	Promoting professional learning for continuous improvement
IV	Facilitating improvements in instruction and student learning
V	Promoting the use of assessments and data for school and district improvement.
VI	Improving outreach and collaboration with families and community
VII	Advocating for student learning and the profession

Table 2.2 Descriptive examples of the SEF mapped to the domains of the Teacher Leader Model Standards

Domain	Description	Descriptive examples from the Science Education Fellowship (SEF)
I	Fostering a collaborative culture to support educator development and student learning	For the entire first year of SEF, teachers work in collaborative groups with the goal of improving their own practice as well as student learning. Together, they plan, enact, and review lessons centered around disciplinary core ideas and science and engineering practices.
II	Accessing and using research to improve practice and student learning	Research from the field informs strategies that employ in their own classrooms as well as the reflective conversations within their collaborative teams. All courses of study in Year 1 of the program have research articles that guide the work.
III	Promoting professional learning for continuous improvement	SEF teachers have demonstrated that they are committed to professional learning for both themselves and for other teachers, in all stages of their Fellowship and as alumni.
IV	Facilitating improvements in instruction and student learning	Within the SEF, teachers are committed to improving science instruction and student learning. The teachers review recordings of each other's lessons, focusing on evidence of student learning and effective pedagogical practices.
V	Promoting the use of assessments and data for school and district improvement	An entire monthly meeting in Year I of the SEF focuses on student assessment. As teachers review one another's lesson videos, they also review student work and discuss next steps based on the evidence. As they plan their GPS projects, they consider district needs and goals, including responding to data.
VI	Improving outreach and collaboration with families and community	Many Fellows choose to focus their GPS projects on involving families and community partners to improve science instruction. Fellows note that an important element of teacher leadership is including all members of the school community as valuable partners.
VII	Advocating for student learning and the profession	As teacher leaders, Fellows become strong advocates for effective science instruction, the need for professional learning, needed curriculum changes, and other important ideas. Fellows present at school board meetings, local and national conferences, and lead professional learning opportunities within their schools and districts.

Distributed leadership

SEF prepares teachers for leadership opportunities within a distributed leadership model. Distributed leadership models have postulated that limiting leadership to a set of leader traits and characteristics does not sufficiently explain the "how" and the "why" of working systems (Spillane et al., 2001). The "how" and "why" can only be explained by looking at the leaders and their followers and the context in which their interactions take place. In school districts, the formal leaders typically include superintendents, assistant superintendents, principals and curriculum coordinators. One science coordinator is often given the responsibility of supporting the entire district's science initiatives and interacting with superintendents, principals and teachers (Whitworth et al., 2017). Some districts have not had these positions, leaving science coordination to department chairs at the elementary, middle, or high schools. The distributed leadership model expands the number of people who guide a district. Specifically, teachers are given more voice by providing them with the knowledge and skills needed to collaborate with other district leaders.

A unique framework for investigating distributed leadership within high schools has been developed by Halverson and Clifford (2013). Their work extends beyond simply how factors such as policies, external environment, and situational analysis influence leadership practices. Instead, they conceptually divide the school into two different environments for the work of leaders – the leading environment and the learning environment. "Analyses of the leadership environment portray how school contexts enable, constrain, and afford leadership action" (Halverson & Clifford, 2013, p. 391). This environment includes the more traditional facets of leadership such as budgets, resource planning, hiring, and facilities management. Success in the leading environment provides for a learning environment conducive to success for all students. Malin and Hackmann (2017) provide an interesting case study of an urban high school utilizing distributed leadership to move from a traditional high school structure to a college and career academy model to improve engagement of students with the intention of better preparing them for either college or the workplace. In this case, the leadership is distributed externally from the high school and assumed by the community volunteers and committee members. Other studies regarding distributed leadership explore how this orientation impacts the roles that teachers hold (e.g., Camburn et al., 2003) or how distributed leadership is enacted (de Lima, 2008). Lastly, it must be recognized that the process of distributed leadership will rely more heavily on multiple contributors. This can be a barrier if stakeholders won't embrace the work required for change (Freeth et al., 2014).

With a focus on deep science learning and model of distributed leadership within and beyond classrooms, SEF has led to true transformation within many of the districts served. The next section provides more specifics on what the program looks like over the two-year Fellowship and beyond.

Context: National implementation of the Science Education Fellowship (SEF) program

Informed by the literature, including some of the studies discussed above, the SEF is purposefully designed to support teachers of science as they build their capacity as teacher leaders. The design of the fellowship considers what is known about adult learning, effective PD, models of teacher leadership, and exemplary science instruction.

In Year 1 of the program, the selected Fellows participate in 125 hours of PD in which these teachers work on improving their instruction through PLCs (DuFour & Eaker, 1998; Stoll et. al., 2006), tuning protocols (Easton, 2009; McDonald et al., 2013), and lesson study (Cerbi & Kopp, 2006; Rock & Wilson, 2006). This intensive professional learning experience results in a program in which these potential leaders reflect upon their own teaching (guided by research-based methods and the Framework/NGSS), while working collaboratively with colleagues. This year of professional learning moves teachers through the first two boxes in the Theory of Action in Figure 2.1.

In the first semester, PLCs comprise mixed grade-level K-12 teachers, called "vertical teams" and engage in 125 hours of professional learning. Each team chooses a course of study including a Disciplinary Core Idea (DCI) (Frameworks, NRC; NGSS) and a teaching practice or method supported by a research article. In the second semester, the teachers form "horizontal teams" across similar grade bands. Their course of study includes one Science and Engineering Practice (SEP) (Frameworks, NRC; NGSS) and a different pedagogical strategy or method supported by a research article. These two courses of study become the lens through which the Fellows analyze each other's lessons. The PLCs function as collaborative coaching and learning in science (CCLS) teams where they provide feedback on videos of each other's lessons, the lesson plans, and student artifacts (Chen et al., 2014).

When not working in their CCLS teams, Fellows participate collaboratively across grade levels and with teachers from other districts to explore varied topics for PD, including engineering design approaches, mandated state exams, presentation skills, data-driven decision-making and what it means to be a teacher leader, among others. They also meet with their respective science coordinators to learn about science education priorities in their own school districts.

In Year 2 of the program, the teachers engage in an additional 125 hours of professional learning. A signature experience of this year is the creation and enactment of a PD plan including a leadership project that is referred to as a GPS[1]. The GPS has two components: (1) 50 hours of support for district initiatives; and (2) 50 additional hours for a personal professional learning plan focused on improving practice. The district-aligned goal is chosen in consultation with the district science coordinator. The GPS is approved by the district coordinator and a mentor assigned by the university who meets regularly with the Fellow. In addition, there are quarterly cohort meetings with the other Fellows in their cohort. It is through their actions in Year 2 that their leadership knowledge and practices are operationalized and refined so that the SEF Fellows are supporting institutional change. This year of professional learning moves the teachers through the next set of boxes in Figure 2.1, which are increased teacher learning and change in teacher practice.

The GPS projects will be discussed in more detail in Chapter 4. Suffice it to say that each Fellow completes a project that is chosen by the Fellow and reflects their views of what is important for them and their colleagues regarding professional learning. In each district, four Fellows per year implement their GPS. Over the four years, 12 Fellows have moved themselves and their schools forward. These projects take place in elementary, middle, and high schools. During a four-year period, 60 Fellows across approximately five districts will each complete a GPS at each site.

Past project evaluation

The evidence shows that the SEF can create lasting changes in instructional practices for the teachers who participate, and that these teachers become leaders within their school districts. David Heil and Associates, Inc. (DHA) and Anne Gurnee have served as the external evaluators for the SEF program since 2014. They have used a mixed method, cross-site approach to gather both qualitative and quantitative data to assess the effectiveness of the SEF program in achieving its stated goals, which include changes in teacher practice, use of research to guide and evaluate lessons, and developing a cadre of teacher leaders in each partnering district. Each year, the evaluators tracked the Fellows involved in the program to assess the impact on their professional growth and have also gathered data from program administrators at the universities as well as district administrators to gauge the fidelity of the program's implementation across the sites and the program's effect on the universities, districts and schools involved.

There has been broad agreement among the involved key stakeholders (e.g., Fellows, District Coordinators, higher education faculty and staff) that the SEF has been successfully implemented in each of the seven regions. The data from this evaluation study indicate that the program has been positively received by participants and has resulted in numerous outcomes and impacts for Fellows, District Coordinators, and their associated schools and districts. Findings from the evaluation study have indicated that:

- There is continued evidence that Fellows involved in the SEF improve their instructional practices, science content knowledge, reflective practice, and use of research in teaching practices. Many Fellows report that the program is a transformational experience for them professionally.
- Fellows have increased confidence in leadership practices such as giving/receiving feedback, informally supporting fellow teachers, presentation skills, and leading professional learning experiences in their schools/districts.
- One of the hallmark features of the program is the creation of a network of professionals who support each other, share knowledge, improve instruction, and ultimately effect change in their schools and districts. Participants value their engagement with this network and want to see it continue over time.

The SEF program has given participating universities a unique opportunity to analyze the experiences of teachers as they develop into teacher leaders through research at the early implementation sites (e.g., Gunning et al., 2020; Hillman, 2016; Klein et al., 2018; Marrero, 2016). Rahman et al., 2018). These findings complement the evaluation research conducted by DHA.

As school districts and universities consider implementing a SEF program of their own, it is worth noting that the program has been run across the United States and in school districts of varying sizes and with different student populations. Table 2.3 illustrates that the SEF program has been implemented successfully in large and small districts; in urban, suburban, and rural communities; and in high poverty locales.

Conclusion

In this chapter, we have presented a vision of how science teacher leadership can drive and enable district transformation. The professional learning and leadership opportunities afforded to the Fellows are beneficial for the school district, the individual teachers,

Table 2.3 Demographics of participating districts[2,3]

	District 1		District 2		District 3		District 4		District 5	
	Enrollment	%FRL	Enrollment	%FRL	Enrollment	%FRL	Enrollment	%FRL	Enrollment	%FRL
CA	8,000	29	4,800	28	5,100	35	52,600	55	28,700	46
FL	220,000	59	75,000	54	101,000	62				
MA	51,400	76	5,800	39	7,000	48	6,564	68	2,800	18
MO	1,500	78	17,000	46	295	53	2,300	50	1,400	35
Districts 6 and 7:			3,800	59	692	70				
NJ	5,400	56	10,600	56	6,700	16	5,000	75	3,800	6
NY	6,600	56	4,600	79	2,700	54	10,700	84	9,700	58
TX	29,000	75	33,500	75	7,500	84	9,700	73	7,900	69

and of course the students. Through participation in the SEF, each school district will have more capable, trained, and confident leaders to enact positive change for student achievement. The teachers will have new skills and a fulfilling experience, leading to meaningful change on a scale larger than their own classrooms.

Notes

1 GPS projects in the past have included: Integrating Biotechnology and Engineering into the Biology Curriculum; Astronomy Club: What's Out There?; The Science Expo: A Children's Book About ELL Students and Science; Making Thinking Visible Through STEM and Science Talk; Woodworking – A Year-long Kindergarten Experience; and Technology Fusion: Incorporating "low tech" and "high tech" experiences into the Biology classroom.
2 Where Free and Reduced Lunch (FRL) eligibility is not reported, we have used the reported % for "economically disadvantaged" or "socioeconomically disadvantaged."
3 Note that MO is working with 7 districts and FL is working with only 3 districts (owing to the size of districts).

References

Banilower, E.R. (2019). Understanding the big picture for science teacher education: The 2018 NSSME+. *Journal of Science Teacher Education*, *30*(3), 201–208. https://doi.org/10.1080/1046560X.2019.1591920

Barth, S. G., Lotter, C., Yow, J. A., Irdam, G., & Ratliff, B. (2023). Understanding the process of teacher leadership identity development. *International Journal of Leadership in Education*, 1–28. https://doi.org/10.1080/13603124.2023.2258083

Camburn, E., Rowan, B., & Taylor, J.E. (2003). Distributed leadership in schools: The case of elementary schools adopting comprehensive reform models. *Educational Evaluation and Policy Analysis*, *25*(4), 347–373.

Cerbin, W., & Kopp, B. (2006). Lesson study as a model for building pedagogical knowledge and improving teaching. *International Journal of Teaching and Learning in Higher Education*, *18*(3), 250–257.

Chen, R. F., Scheff, A., Fields, E., Pelletier, P., & Faux, R. (2014). Mapping Energy in the Boston Public Schools Curriculum. In R. Chen et al. (Eds.), *Teaching and Learning of Energy in K–12 Education* (pp. 135–152). Springer.

Childs-Bowen, D. Moller, G., & Scrivner, J. (2000). Principals: Leaders of leaders. *NASSP Bulletin, 84*(616), 27–34.

Cosenza, M. N. (2015). Defining teacher leadership affirming the teacher leader model standards. *Issues in Teacher Education, 24*(2), 79–99.

Criswell, B. A., Rushton, G. T., Nachtigall, D., Staggs, S., Alemdar, M., & Cappelli, C. J. (2018). Strengthening the vision: Examining the understanding of a framework for teacher leadership development by experienced science teachers. *Science Education, 102*(6), 1265–1287. https://doi.org/10.1002/sce.21472

de Lima, J. A. (2008). Department networks and distributed leadership in schools. *School Leadership and Management, 28*(2), 159–187.

Desimone, L. M. (2009). Improving impact studies of teachers' professional development: Toward better conceptualizations and measures. *Educational Researcher, 38*(3), 181–199. https://doi.org/10.3102/0013189X08331140

DuFour, R., & Eaker, R. (1998). *Professional learning communities at work: Best practices for enhancing student achievement.* Solution Tree.

Easton, L. B. (2009). *Protocols for Professional Learning.* The Professional Learning Community Series. ASCD.

Freeth, W., Andreotti, V. dO., & Quinlivan, K. (2014). Reconceptualizing leadership in the implementation of the New Zealand Curriculum: Implications for school leaders. *International Journal of Leadership in Education, 17*(1), 83–102.

Galosy, J., Mohan, L., Mohan, A., Miller, B., & Bintz, J. (2017). *Math and STEM teacher leadership development: Findings from research and program reviews* (No. 2017-03). Research Report.

Gess-Newsome, J., Luft, J. A., & Bell, R. L. (2009). *Reforming secondary science instruction.* NSTA Press.

Gunning, A. M., Marrero, M. E., Hillman, P. C., & Brandon, L. T. (2020). How K-12 teachers of science experience a vertically articulated professional learning community. *Journal of Science Teacher Education, 31*(6), 705–718. https://doi.org/10.1080/1046560X.2020.1758419

Halverson, R., & Clifford, M. (2013). Distributed instructional leadership in high schools. *Journal of School Leadership, 23*(March), 389–419.

Harris, A. (2003). Teacher leadership as distributed leadership: Heresy, fantasy or possibility? *School Leadership & Management, 23*(3), 313–324. https://doi.org/10.1080/1363243032000112801

Hillman, P. C. (2018). *Vertically aligned professional learning communities as a keystone for elementary science teacher professional development, growth, and support.* Teachers College, Columbia University.

Hofstein, A., Carmeli, M., & Shore, R. (2004). The professional development of high school chemistry coordinators. *Journal of Science Teacher Education, 15*(1), 3–24.

Holland, J. M., Eckert, J., & Allen, M. M. (2014). From preservice to teacher leadership: Meeting the future in educator preparation. *Action in Teacher Education, 36*(5–6), 433–445. https://doi.org/10.1080/01626620.2014.977738

Howe, A. C., & Stubbs, H. S. (2003). From science teacher to teacher leader: Leadership development as meaning making in a community of practice. *Science Education*, *87*(2), 281–297.

Hutchinson, A. E., Schaefer, J., Zhao, W., & Criswell, B. (2023). Practice, tensions, and identity: Boundary crossing and identity development in urban teacher leaders. *Journal of Research on Leadership Education*, *18*(4), 575–599. https://doi.org/10.1177/19427751221113368

Klein, E. J., Taylor, M., Munakata, M., Trabona, K., Rahman, Z., & McManus, J. (2018). Navigating teacher leaders' complex relationships using a distributed leadership framework. *Teacher Education Quarterly*, *45*(2), 89–112. https://www.jstor.org/stable/90020316

Lotter, C., Yow, J. A., Lee, M., Zeis, J. G., & Irvin, M. J. (2020). Rural teacher leadership in science and mathematics. *School Science and Mathematics*, *120*(1), 29–44.

Luft, J. A., & Hewson, P. W. (2014). Research on teacher professional development programs in science. In *Handbook of Research on Science Education, Volume II* (pp. 903–924). Routledge.

Luft, J. A., Wong, S., & Ortega, I. (2009). *The National Science Teachers Association state of science education survey*. National Science Teachers Association.

Malin, J. R., & Hackmann, D. G. (2017). Enhancing students' transitions to college and careers: A case study of distributed leadership practice in supporting a high school career academy model. *Leadership and Policy in Schools*, *16*(1), 54–79.

Marrero, M. E. (2016). Core competencies and strategies for effective leadership. *Global Education Review*, *3*(2).

McDonald, J. P., Mohr, N., Dichter, A., Donald, E. C., & Lieberman, A. (2013). *The power of protocols: An educator's guide to better practice* (3rd ed.). Teachers College Press.

NASEM. (2015). *Science teachers' learning: Enhancing opportunities, creating supportive contexts*. National Academies Press.

Natale, C., Gaddis, L., Bassett, K., & McKnight, K. (2013). Creating Sustainable Teacher. Career Pathways: A 21st Century Imperative, a joint publication of Pearson & National Network of State Teachers of the Year.

National Research Council [NRC]. (2012). *A framework for K-12 science education: Practices, crosscutting concepts, and core ideas*. National Academies Press.

Rahman, Z. G., Munakata, M., Klein, E., Taylor, M., & Trabona, K. (2018). Growing our own: Fostering teacher leadership in K-12 science teachers through school-university partnerships. In J. Hunzicker (Ed.), *Teacher leadership in professional development schools* (pp. 235–253). Emerald Publishing Limited.

Rock, T. C., & Wilson, C. (2005). Improving teaching through lesson study. *Teacher Education and Quarterly*, *32*(1), 77–92.

Sinha, S., & Hanscin, D. L. (2017). Development of teacher leader identity: A multiple case study. *Teaching and Teacher Education*, *63*(2017), 356–371. https://doi.org/10.1016/j.tate.2017.01.004.

Spillane, J. P., Halverson, R., & Diamond, J. B. (2001). Investigating school leadership practice: A distributed perspective. *Educational Researcher*, *30*(April), 23–28.

Stoll, L., Bolam, R., McMahon, A., Wallace, M., & Thomas, S. (2006). Professional learning communities: A review of the literature. *Journal of Educational Change*, *7*(4), 221–258.

Teacher Leader Model Standards. (2011). Retrieved 17 Feb 2023 from https://www.ets.org/content/dam/ets-org/pdfs/patl/patl-teacher-leader-model-standards.pdf

Wenner, J. A., & Campbell, T. (2017). The theoretical and empirical basis of teacher leadership: A review of the literature. *Review of Educational Research*, *87*(1), 134–171. https://doi.org/10.3102/0034654316653478

Whitworth, B. A., & Chiu, J. L. (2015). Professional development and teacher change: The missing leadership link. *Journal of Science Teacher Education*, *26*, 121–137. https://doi.org/10.1007/s10972-014-9411-2

Whitworth, B. A., Maeng, J. L., Wheeler, L. B., & Chiu, J. L. (2017). Investigating the role of a district science coordinator. *Journal of Research in Science Teaching*, *54*, 914–936. https://doi.org/10.1002/tea.21391

Whitworth, B. A., Wenner, J., & Tubin, D. (2022). Science teacher leadership: The current landscape and paths forward. In J. Luft & M. G. Jones (Eds.), *The handbook of research on science teacher education* (pp. 257–272). Taylor & Francis.

York-Barr, J., & Duke, K. (2004). What do we know about teacher leadership? Findings from two decades of scholarship. *Review of Educational Research*, *74*(3), 255–316.

Yow, J. A., Wilkerson, A., & Gay, C. (2021). Mathematics and science teacher leadership understanding through a teacher leadership course. *International Journal of Science & Mathematics Education*, *19*(4), 839–862. https://doi.org/10.1007/s10763-020-10080-y

Chapter 3

Collaborative Coaching and Learning Science (CCLS)

Arthur Eisenkraft, Amanda Gunning, and Meghan Marrero

Working with teachers from different grade bands, such as collaborating with elementary and high school teachers, provided valuable support for meeting the needs of my students. This collaboration allowed for the sharing of common goals, the selection of the same Science and Engineering Practices (SEPs) and content standards, and the exploration of how content can be taught across grade bands. The opportunity to observe how students engage in science throughout their academic careers provides significant value through this atypical approach.

This collaborative approach led to a deeper understanding of how to effectively teach and engage students across different age groups and educational levels. It also allowed for the sharing of best practices and innovative teaching methods that can benefit students at various stages of their academic journey.

Overall, the experience of collaborating with teachers from different grade bands can be enriching and beneficial for both educators and students, fostering a more holistic and comprehensive approach to education.

Nicole Holman, High school, Hillsborough, FL.

The Wipro experience provides teachers with many gifts of collegiality and a shared learning experience. Meeting with middle and high school teachers provided a unique opportunity to

DOI: 10.4324/9781003490586-4

see the future, learning from each particular strand. It was a fascinating opportunity to visit different neighborhoods, observe the culture of various schools, and hear directly from the teachers working in that building and neighborhood. My colleagues blew me away with the depth and rigor of their projects. Connecting the learning dots from elementary to high school illustrated the meaningful interconnectedness of our work.

Karma Paoletti, Elementary School, Cambridge, MA.

I see how our work in elementary school evolves into what the students will see in high school. Working with K-12 Fellows gave me an understanding of science in a broader context.

My experience in the V-CCLS presentation this year has been informative, eye-opening, motivating, and inspirational. I have learned about some processes and programs currently available for improving science education in my classroom and district.

Rick Anderson, High School, TX

The V-CCLS specifically helped me realize and understand exactly how related curriculum is on all levels. The rigor is on different levels, but it is all tiered and built upon over the academic year of the children. I also learned a lot about myself through this process. I have struggled with my confidence as a teacher since I started. I regarded all my V-CCLS group members very highly. When it came down to my debrief video, I was very concerned because of my own self criticalness. What I heard at that debrief was better than expected. I heard so many positive things about my teaching style and methods. I was astonished, relieved, and honored to hear that my group members who I had the utmost respect for saw goodness in what I do. That alone was probably the best experience of the whole semester.

Matthew Gaines, Elementary School, TX

Teaching excellence is at the heart of all professional learning. The first year of the Science Education Fellowship (SEF) program, Fellows focus primarily on the first of the three SEF pillars – reflective practice. They share and critique each other's lessons. They experience and learn about the vertical articulation of K-12 content curriculum in the first half of the year and have an in-depth look at one Science and Engineering Practice (SEP) in the second half of the year. Using protocols, the Fellows build trust and grow professionally. Through that focus, they will gain an understanding of the other two pillars – adult learning and leadership.

We use a Collaborative Coaching and Learning Science (CCLS) model throughout the year. For the vertical articulation of curriculum (V-CCLS), the teams are comprised of teachers from elementary, middle and high schools. For the work on science and engineering practices (H-CCLS), the teams are comprised of grade level teams (e.g., one team of elementary teachers and another team of high school teachers).

The CCLS process begins with the establishment of teams. These teams develop a culture of trust and respect by setting group norms and building a common set of expectations on how they will work together. They then determine their course of study (CoS) and establish a meeting schedule where each Fellow will have the opportunity to teach and share a recording of their lesson as well as student artifacts related to that CoS. They complete one "observation cycle" for each teacher on the team. Finally, they

Table 3.1 An outline of the CCLS structure

I. Establish the CCLS team	II. Engage in CCLS cycle	III. Synthesize learning from CCLS
Create a culture: • Set norms • Develop a common language Determine course of study: • Common learning concept • Research article that will be the lens for observations Establish observation schedule	Complete observation cycles: • Record lesson and share video and pre-observation form with group • Group watches video prior to debrief • Debrief video as a group • Reflection *# of teachers = # of observations*	Reflect on CCLS work and accomplishments: • Revisit the course of study • Teacher reflections • Teacher being observed reflection
Discuss research article to develop a common understanding Structure the archive binder (see notes on SEF site)	Inquire: Be sure to use research article to inform debrief Connect to student learning by looking at student work	Complete the archive binder Share success through presentation and celebrate

hold a synthesizing conversation meeting – and then the development of their archive and presentation about their team's learning.

Table 3.1 provides an outline of the CCLS structure. The rest of this chapter will provide details and examples of each component.

First half of the year – V-CCLS

During the first year of the Fellowship, teachers work together in cross-district groups sharing and critiquing classroom lessons. In the first half of the year, Fellows from different grade bands select a content area (i.e., a disciplinary core idea as defined by the Framework/derivative Next Generation Science Standards [NGSS]) to focus on for their lessons as well as one research article that explores the pedagogical approach they will reflect on in their lessons. During the second half of the year, Fellows from the same grade band select one SEP (from the Framework/NGSS) and a new teaching approach from a research article to try in their lessons. During the whole year, the entire cohort meets monthly with the university leaders and the District Science Coordinators (DSCs). These meetings provide progress checks, sharing of challenges and successes and reflections of lessons learned. Different sites also choose areas of focus for parts of the meeting including discussions of teacher leadership, equity issues, uses of technology in the classroom, or a host of other topics pertinent to the districts and their teachers. These are often determined collaboratively with DSCs (and teachers).

Toward the end of each semester, monthly meetings are used to explain the requirements for the conference, the portfolio, and plans for Year 2 of the Fellowship program.

Working with teachers from different grade bands was challenging, but we were dedicated to holding regular meetings, having open discussions, and collaborating on planning. We talked in-depth about how each grade level contributed and scaffolded concepts to prepare

A Roadmap for Transformative Science Teacher Leadership

students for the next level. Our journey deepened our understanding of the standards and how they fit together to build student knowledge. It helped us develop and practice pedagogical strategies and fostered a spirit of cooperation as a vertical team.

Maria Louisa Soto, Upper elementary (3-5), Arlington Independent School District, TX.

It is the vertical articulation of science concepts during the first semester that sets the SEF apart from many other professional development programs. This is not a quick, cursory review of a district curriculum map where grade topics are listed. Rather, it is an in-depth comparison of the same content lesson taught at multiple grades (i.e., kindergarten, 4th grade, 8th grade, and 11th grade). Rarely do teachers devote more than 50 hours in a semester observing each other's lessons and providing feedback. We call this approach vertical Collaborative Coaching and Learning in Science (V-CCLS). It is the first step in the SEF program and begins immediately following the welcoming ceremony. This work utilizes groups made up of elementary, middle, and high school teachers working cooperatively to better understand the vertical articulation of disciplinary concepts. Vertical articulation of curriculum allows Fellows to better understand the journey of their students from pre-K to high school. From our evaluation, we have learned that this vertical work is novel for most teachers, and some teachers have actually called it, "mind blowing" to have the opportunity to see a science concept worked through the full K-12 spectrum. It helps teachers to have a strong sense of their district's science instructional sequence, including how one concept builds on another and how each topic will be introduced, explored and explained across students' academic trajectory.

For the V-CCLS, the cohort of 20 teachers from five districts are placed into four teams – biology, chemistry, physics, and earth/environmental science. The elementary teachers and some middle school teachers can be placed on any team while the secondary teachers are placed within the discipline matching their certification area. This design is possible because one criterion for the selection of the cohort of 20 was to have an equal distribution of teachers across grade levels and high school teachers of each discipline. An attempt is made to allow for the generalist teachers to be placed in content areas that they are currently or will be teaching that year. Thus, one team might include a 12th grade AP Physics teacher, a 7th grade physical science teacher, a 4th grade teacher, a 2nd grade teacher, and a kindergarten teacher. Another group could consist of a 9th grade biology teacher, a 6th grade life science teacher, a 3rd grade teacher, and a 2nd grade teacher. An example of teams is shared in Table 3.2.

Table 3.2 Example V-CCLS team membership of the cohort of 20 SEF Fellows, where each x represents one teacher

	Biology	*Chemistry*	*Physics*	*Earth science*
Elementary	xx	x	xx	xxx
Middle school	x	xx	xx	x
High school	xx	xx	x	x

To begin the V-CCLS, teachers are matched into vertical groups based on the four content strands of science that they teach noted above. Once teams have been assembled, they meet to define team norms, plan meeting times, and choose a course of study (CoS), which will be the basis of the team's work for the first semester[1] of the program.

The course of study (CoS)

After meeting their V-CCLS team members, the team decides on a course of study for the semester. The course of study has two components: a content topic (referred to as a Disciplinary Core Idea in the Framework and NGSS) and pedagogical approach based on a research article explaining a teaching method. Groups are encouraged to try new methods for this work. Each teacher will teach a lesson on that content topic during the semester appropriate to their grade level while applying the approach explained in the research article.

Prior to planning these lessons, teachers endeavor to consider this content more deeply through a process of Curriculum Topic Study (CTS) (Keeley, 2005). CTS (or other similar strategies) allows "K-12 educators to deepen their understanding of the important science and mathematics topics they teach. CTS builds a bridge between state and national standards, research on students' ideas in science and mathematics, and opportunities for students to learn ... through improved teacher practice" (Mundry et al., 2009, p. viii). In summary, CTS is a process that allows teachers to deeply explore content through professional development that supports them in examining adult level content related to the topic, common misconceptions, effective pedagogical strategies, and connections to other topics. Through the CTS process, teachers can improve their own content knowledge and skill in teaching science (Mundry et al., 2009). Teachers continue this CTS throughout the CCLS process.

Together, team members examine the NGSS or state standards for their science content area and choose a content topic from within the NGSS Disciplinary Core Ideas (DCI) to study. For instance, the chemistry team might examine PS1.A, Structure and properties of matter. Ultimately, each teacher will teach a lesson in which students explore an NGSS performance expectation related to this topic. Table 3.3 shows sample performance expectations at each grade band for PS1.A. Table 3.3 illustrates an example of how student content knowledge, use of SEPs, and understanding of crosscutting concepts should progress in their studies of science from elementary through high school.

Table 3.3 Example NGSS performance expectations across grade levels

Primary (2nd grade)	2-PS1-1 Plan and conduct an investigation to describe and classify different kinds of materials by their observable properties.
Upper elementary (5th grade)	5-PS1-1 Develop a model to describe that matter is made of particles too small to be seen.
Middle school	MS-PS1-1 Develop models to describe the atomic composition of simple molecules and extended structures.
High school	HS-PS1-1 Use the periodic table as a model to predict the relative properties of elements based on the patterns of electrons in the outermost energy level of atoms.

A Roadmap for Transformative Science Teacher Leadership

The second part of the CoS is a pedagogical approach grounded in a research article that the team selects. The decision of which instructional method will be the focus of the lessons requires the team members to identify a research article that underpins the method that interests them all and that may address challenges in the development of conceptual understanding of the particular performance expectation of the content topic. To select the article and method, they read research articles, take notes, and discuss and make a decision as a team on one article and its instructional strategy. Since teachers don't generally have access to many research journals, the university overseeing the program provides support. For example, the teachers may find an article on Google Scholar and the university can access that article for them. University partners may also provide a folder of sample articles to choose from or to serve as examples of what would be an appropriate article.

The selection of a research-driven instructional method is important because it pushes teachers to review research articles on varied instructional methods. This research review is something most participating teachers have not done recently and is a valuable professional activity. Reading research articles as a practicing teacher is a very different experience from reading similar articles while a graduate student. Utilizing research is one of the domains of the teacher-leader model standards discussed in Chapter 2.

Research information has given a new window to observe how my students learn in my classroom and I am able to help them better by knowing what they know and how I can help grow when working in groups. Our research helped me to shift the focus of Assessment and Feedback from being mostly to help me as the instructor to make decisions, to focusing on how Assessment and Feedback can inform the students and guide them further in their learning. (Survey Respondent 2, CA, March 2019)

The information I get from the research pretty much reintroduced me to best practices that should be done in the classroom. The thing that was different about this paper is that it actually provided data to support their ideas. (Survey Respondent 3, CA, March 2019)

Teachers are challenged to explore a CoS pedagogy that focuses on a method of instruction they have not yet tried or are inexperienced using. The range of conversations that you can have around your CoS is limitless. To promote professional growth and success, the selection of the CoS in both the vertical and horizontal CCLSs should explore instructional methods and strategies that are new or have been challenging for the teachers on the team, and the conversations need to focus on student learning in science. According to DuFour, the mission "is not simply to ensure that students are taught but to ensure that they learn. This simple shift –from a focus on teaching to a focus on learning – has profound implications."[2] (Hord, 1997; Hord & Sommers, 2008).

The CoS pedagogy must focus on science instruction but is otherwise up to the group to choose. Typically, each teacher will bring one or two articles to the group for consideration and the CoS pedagogy is selected by consensus. Examples of CoS teachers have chosen for their CCLS work includes inquiry-based learning; Total Physical Response;

Table 3.4 Example CoS for one cohort of fellows

V-CCLS team	Course of Study (CoS)		Citation
	Content topic	Research-based method of instruction	
Biology	Ecosystems	Using cooperative learning	Rabgay, T. (2017). The effect of using cooperative learning method on tenth grade students' learning achievement and attitude towards biology. *International Journal of Instruction, 11*(2), 265–280.
Chemistry	Structure and properties of matter	Modeling in chemistry	Posthuma-Adams, E. (2014). How the chemistry modeling curriculum engages students in seven science practices outlined by the College Board. *Journal of Chemical Education, 91*(9), 1284–1290.
Physics	Energy	Argumentation in the science classroom	O'Connor, C., Michaels, S., & Chaplin, S. (2015). 'Scaling Down' to explore the role of talk in learning: From district intervention to controlled classroom study. Retrieved from https://bit.ly/2BRJcrh
Earth science	Water and its impact on landforms and human population	Project-based learning	Chun, M-S., Il Kang, K., Kim, Y. H., & Kim, Y. M. (2015). Theme-based project learning: Design and application of convergent science experiments. *Universal Journal of Educational Research, 3*(11), 937–942.

using analogies; or infusing reading strategies. This method and related research article form the guiding lens for the team's work. Once the team members have agreed on a shared CoS including the content topic (DCI) and pedagogy, they are ready to plan lessons. For example, an earth/environmental science team might discuss how they can each combine the content topic of Earth's place in the universe with the use of simulations in the classroom. Teachers collaboratively discuss their ideas, and consider what the evidence of student learning will be. For example, will we hear student conversations? Read student writing? Examine student models? Tables 3.4 and 3.5 summarize some example CoS that previous Fellows have followed. These examples show the grain size of the content topic and the different pedagogical approaches that were selected.

Once the lessons are planned, each teacher implements their lesson with students and video records the classroom to be viewed by their team members. This lesson now benefits from an in-depth study of both the content and pedagogical approach that the Fellow may not have had previously. After recording their lesson, each teacher shares the final lesson plan, video, examples of student work, and a short introductory sheet with the rest of the team. The introductory sheet provides some contextual information such as the type of class, prior lessons, and specific things that the teachers want their teammates to focus on for feedback.

It was powerful working with middle school and high school teachers. As an elementary teacher, it helps to know what prior knowledge our students will need before they get to those

Table 3.5 Example CoS for a second cohort of SEF fellows

| V-CCLS team | Course of Study (CoS) | | Citation |
	Content topic	Research-based method of instruction	
Biology	Energy transfer in ecology	Place-based education in science	Autreau, B. T., & Binns, I. C. (2012). Investigating student attitudes and achievements in an environmental place-based inquiry in secondary classrooms. *International Journal of Environmental and Science Education, 7*(2), 167–195.
Chemistry	Chemical reactions	Phenomenon-based learning	Wakil, K., Rahman, R., Hasan, D., Mahmood, P., & Jalal, T. (2019). Phenomenon-based learning for teaching ICT subject through other subjects in primary schools. *Journal of Computer and Education Research, 7*(13), 205–212. https://doi.org/10.18009/jcer.553507
Physics	Energy	Science notebooking	Sparks, B. M. (2016). The effect of inquiry with science notebooks on student engagement and achievement, Professional paper submitted for a Masters' degree in Science Education, Montana State University, Bozeman, Montana.
Earth science	Earth's changing surface	Improving student achievement using notebooks	Shelton, A., Smith, A., Wiebe, E., Behrle, C., Sirkin, R., & Lester, J. (2016). Drawing and writing in digital science notebooks: Sources of formative assessment data. *Journal of Science Education and Technology, 25*, 474–488. https://doi.org/10.1007/s10956-016-9607-7

grades and how the standards in elementary school are interconnected with those in middle and high school. It was enlightening to see how this alignment can benefit our students.

By sharing insights into what each grade band covers, we all gained a better understanding of the bigger picture of our students' educational journey. It reminded us that things are not always as they seem.

Janine Hogel, Upper elementary (3-5), Clifton, NJ.

Live debriefing

Team members view their colleagues' lesson videos independently, taking notes on what they see and hear happening in the classrooms, using provided forms. Next, the team comes together, ideally in person or virtually, if necessary, to debrief one of the Fellows' lessons. These debriefs are spread over the semester. The team may meet every three or four weeks to debrief a single lesson. Each teacher's lesson gets its own debrief time, which is structured using a tuning protocol (adapted from Allen & McDonald, 2003) and described further here. Each debrief is facilitated by a different team member whose lesson is not being discussed in the particular session. First, the teacher being observed shares any additional background or context about the lesson and answers any clarifying questions teammates may have (e.g., do the students always work in the same groups?). Next, each team member provides the teacher being observed with a round of "warm feedback" and then followed with a round of "cool feedback" on the lesson. Fellows were given the opportunity to practice rounds of warm and cool feedback in a monthly meeting prior to using it on their own in their teams. Through warm feedback, observers share interesting or positive aspects of the lesson. For instance, "I noticed that your students seamlessly integrate academic vocabulary" or "I appreciated that you used a lot of strategies to promote student discourse during the science investigation." Cool feedback is not negative feedback. Rather, observers share different perspectives, ideas, or wonderings. For example, "I wonder whether you considered having students jot down their ideas individually prior to the class discussion," or "I would have liked to see more opportunities for students to support their ideas with evidence on the lab handout." Our evaluation has revealed how transformative this feedback process can be for the Fellows. Although they find it difficult at first to give feedback, most come to appreciate this new skill greatly, and nearly all Fellows welcome the feedback from their peers. Student artifacts are an integral part of the protocol. This gives everyone a chance to see what the students are learning. This is another key distinguishing aspect of the SEF. The team viewing of these videos is not simply a performance by the teacher on video. The lesson is reviewed wholistically by a group of colleagues with similar intentions – improved science instruction and learning. Both warm and cool feedback must be specific and supportive and should not include general statements. Warm feedback precedes cool feedback. The warm feedback reinforces the notion that the observers are supporting the teacher and are there to help. This sets the stage for the teacher to lower their anxiety level and listen more intently to the cool feedback. During the sharing of feedback, the teacher being observed does not respond; they listen intently to the feedback and take notes. With the restriction that the teacher being observed remains silent, all of their attention is focused on hearing the feedback. Too often when we are supposed to listen, we have a tendency to simultaneously be constructing responses (i.e., defensive postures). By removing the opportunity to respond, the teacher listens and records the comments and is better able to hear and internalize what is being said. The teachers providing warm and cool feedback should not have an "opt out" option. Everyone should be required to give feedback. Not allowing, "I have no cool feedback" or "my cool feedback is like someone else's" helps generate more open sharing, better communication and a more effective learning environment.

Working with teachers from primary elementary grades through high school gave me more insight into how the science content builds over time. I was able to share a 4th grade lesson on force and motion and later watch an example of a high school lesson on force and motion. Seeing the high school lesson on the same topic helped me to understand how what I teach builds a foundation for students' future learning.

Dawn Avolt, Upper elementary (3-5), Pinellas County Schools, FL.

Working with teachers at different grade levels broadened my knowledge, reassured me that my problems and concerns are not unique, and gave me an opportunity to share my experiences with other educators. As elementary classroom teachers, we spend the vast majority of our time in our own little world. We tend to fall into daily routines and habits of mind. It also is helpful to know what my students learned prior to the grade I teach, and better prepare them for their future experiences.

David Kleiner, Upper elementary, Clifton, NJ

The warm feedback was helpful because it helped me see where teachers agreed with my choices and decisions. It helped me to know that taking risks is okay to push yourself beyond the normal margins from doing the same thing to using curriculum as a springboard to teach content with never-before used skills and materials.

The cold feedback was helpful because it helped me realize that everyone does not have to agree with my work. Cold feedback helps me think objectively concerning my work. Both forms of feedback are effective and critical to one's development as a professional.

Marsha Bolden, High School, Irving ISD, TX.

Following the period of warm and cool feedback, the teacher being observed may respond to the process without rebutting any of the comments. This is important in that the feedback is about the lesson, not personal feedback. At this time, the team also reflects on the lesson in terms of what the experience taught them about both aspects of the course of study – the content topic and pedagogy.

Upon completion of the tuning protocol (i.e., rounds of warm and cool feedback), the teachers discuss student artifacts (e.g., student notebooks, diagrams) from the lesson. During this time, the entire dynamic of the group changes as teachers shift to dissection and analysis of student work. The focus is on student learning and the impact of the teaching. For instance, a team member might note that a student's drawing of a model reflects a common misconception, and the team can discuss ways to follow the lesson to facilitate conceptual change. Teachers are encouraged to refer to their chosen research article, CTS documents, and student work to make connections.

The V-CCLS is repeated every few weeks with a different team member going through this process in turn, providing everyone the chance to give and receive feedback (a key leadership skill) and a chance to moderate the discussion (another key leadership skill).

To ensure that the tuning protocol is followed both in format and time duration, the debrief sessions are recorded.

The tuning protocol

The "tuning protocol" was developed by David Allen and Joe McDonald at the Coalition of Essential Schools primarily for use in looking closely at student exhibitions. Here, it is a process for reflection on teaching and related student work. In the outline featured next, unless otherwise noted, time allotments indicated are the suggested minimum for each task.

I **Introduction** [10 minutes]. Facilitator briefly introduces protocol goals, norms, and agenda. Participants briefly introduce themselves.

II **Teacher Presentation** [20 minutes]. Presenter describes the context for student work (its vision, coaching, scoring rubric, etc.) and presents samples of student work (such as photo- copied pieces of written work or video tapes of an exhibition).

III **Clarifying Questions** [15 minutes maximum]. Facilitator judges if questions more properly belong as warm or cool feedback than as clarifiers.

IV **Pause to reflect on warm and cool feedback** [2–3 minutes maximum]. Participants make note of "warm," supportive feedback and "cool," more distanced comments (generally no more than one of each).

V **Warm and Cool Feedback** [15 minutes]. Participants among themselves share responses to the work and its context; **teacher-presenter is silent**. Facilitator may lend focus by reminding participants of an area of emphasis supplied by teacher-presenter.

Warm:
- Deliberately supportive and specific
- Appreciative perspective

Cool:
- Deliberately challenging perspective
- Suggests what is missing or may need to be developed

VI **Reflection/Response** [15 minutes]. Teacher-presenter reflects on and responds to those comments or questions he or she chooses to. Participants are silent. Facilitator may clarify or lend focus.

VII **Debrief** [10 minutes]. Beginning with the teacher-presenter ("How did the protocol experience compare with what you expected?"), the group discusses any frustrations, misunderstandings, or positive reactions participants have experienced. More general discussion of the tuning protocol may develop.

Guidelines for facilitators

1 Be assertive about keeping time. A protocol that doesn't allow for all the components will do a disservice to the presenter, the work presented, and the participants' understanding of the process. Don't let one participant monopolize.

2 Be protective of teacher-presenters. By making their work more public, teachers are exposing themselves to kinds of critiques they may not be used to. Inappropriate comments or questions should be recast or withdrawn. Try to determine just how "tough" your presenter wants the feedback to be.

3   Be provocative of substantive discourse. Many presenters may be used to blanket praise. Without thoughtful but probing "cool" questions and comments, they won't benefit from the tuning protocol experience. Presenters often say they'd have liked more cool feedback.

The full process for the SEF program is summarized in Table 3.6 (textually) and Figure 3.1 (graphically).

Format of SEF V-CCLS meeting protocol

Table 3.6 Summary of V-CCLS process

Prior to meeting

- **Presenters**
  - Prior to the presentation, presenters fill out "Form 1: V-CCLS Pre-Observation Form."
  - Presenters hand out completed "Form 1: Pre-Observation" to observing Fellows.
  - Presenters present a 20-minute presentation on their V-CCLS work.
  - During warm and cool feedback, presenters take notes on "Form 4: Recording Form." This is for them to keep and include in their binders.
  - Once presentation and feedback are complete, presenters have three minutes to fill out "Form 6: Presenter Reflection Form." This is for them to keep and include in their binders.
- **Observers**
  - Read the pre-observation information (Form 1).
  - Listen to V-CCLS presentation.
  - During the presentation, observers take notes and write down any questions on "Form 2: Observation Notes."
  - After the presentation, observers have three minutes to fill out "Form 3: Feedback Form."
  - Observers provide warm and cool feedback to presenters (eight minutes total).
  - Personal reflection – Observers now have three minutes to reflect on what they have learned from this presentation. Observers fill out "Form 5: Observer Reflection Form." Observers will keep this and include it in their binders.
  - Observers hand "Form 3: Feedback Form" to presenters.
- **Facilitator**
  - Hands out forms to observers (excluding Form 1).
  - Ensures all protocols are followed and are on-time.
  - Brings a watch so that he or she can keep time during the meeting.
- **Required forms**
  - **Form 1, Pre-observation Form** – For presenters to hand out to Fellow observers prior to the presentation.
  - **Form 2, Observation Notes** – For observers to take notes on during the presentation and keep for themselves.
  - **Form 3, Feedback Form** – For observers to write warm and cool feedback and give to presenters.
  - **Form 4, Recording Form** – For presenters to take notes during warm and cool feedback.
  - **Form 5, Observer Reflection Form** – For observers to reflect on what they learned from the presentation.
  - **From 6, Presenter Reflection Form** – For presenters to reflect on what they learned from presenting.

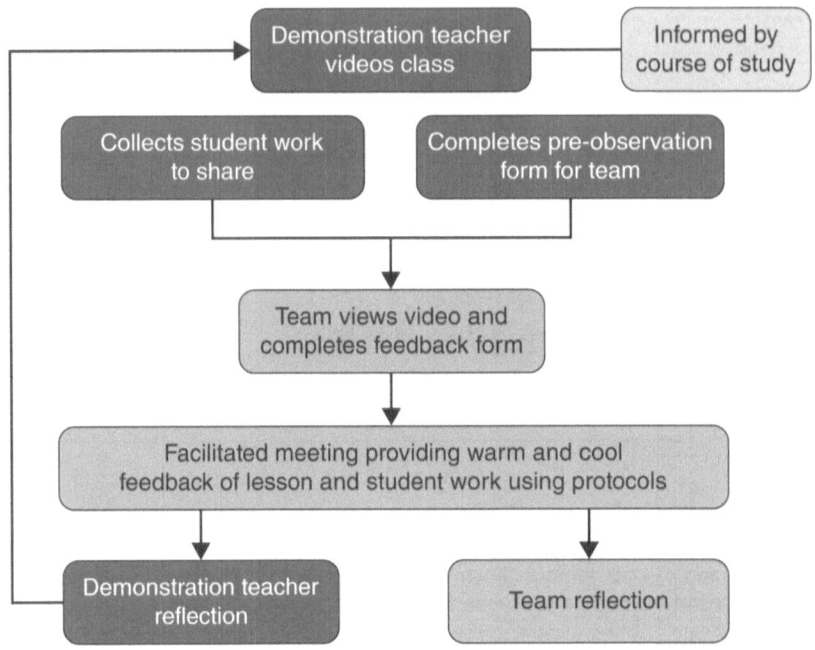

Figure 3.1 Graphical presentation of V-CCLS process.

Sources: Image by the author.

Learning how to give warm and cool feedback has been a valuable resource, both as I grow as a teacher-leader and coach, and for providing my students with feedback on their own work. Learning to ask questions in a non-judgmental way, such as "Why did you decide to do x?" and "What do you think would happen if you did y?" has completely changed how I interact with colleagues and my students.

Regina Borriello, High School, Clifton, NJ.

Conference

When the V-CCLS semester is complete, teachers must prepare to present their work to the whole group. This presentation typically occurs in a large meeting of all the V-CCLS groups. An important part of the presentation preparation will be how teachers must reflect upon the CoS (science content and the pedagogical research article) they began with and the evidence of their own students' learning during the lesson they developed. Teachers discuss these findings among their group to make meaning and identify themes that they may present to the audience (the rest of the cohort, science coordinators, university staff, district admin, etc.). This process is often one that teachers find powerful as they are able to make connections across classrooms on student learning, content, and instructional method. For example, teachers may find that students in all grades continue to exhibit the same science misconceptions on their chosen content topic.

Creating the presentation and ensuring that it involves all team members and meets the time constraint provides an opportunity for additional professional learning.

I found the amount of work and reading hard to manage at the beginning, but as I was able to find my stride, I learned to parse my time more effectively. As with anything that is challenging, I also experienced triumphs. Getting through my first presentation was one of them. It may sound like a small one, but when you feel small, the experience is big. I also found my voice and learned to advocate for myself. This experience bled into my work as a teacher and teacher-leader. I found myself thinking in terms of what kind of feedback I would want and what kind of help I found useful when speaking and working with fellow teachers. I had expected this fellowship to be an opportunity to learn and grow as an educator. In that respect, I wasn't disappointed. I have grown so much and am far more reflective than I had been before and have learned to research pedagogical strategies for myself instead of waiting for my district or principal to decide what I should know. (Maria Soto (TX Fellow))

The final presentation was the apex of the semester. It allowed the cohort to present our findings to each other and the [SEF] group as a whole. It was at times frustrating, tiring, and tedious. However, in the end it was worth all of the effort. One of the great things about the class was that our university leader did a dry run with the groups a month before. This gave us a peek into how the final presentation would be and what changes to make. On presentation day, I was a wreck... I did not want to let my group down. I wanted to make a great impression. Our group was first. It went well. Not as good as I would have liked but it was still a good presentation. However, since I have done it, I know I can do much better the next time. (Myesia Morrison (TX Fellow))

Teams develop formal presentations to be shared across partner districts for their peers. Each team member participates in the design and presentation of their work. These presentations are timed, should include a slideshow as well as a "handout" (may be electronic), and should actively engage the audience to help them better understand some aspect of the CoS. Feedback on presentations by those present is given using the protocol of warm and cool feedback. For many teachers, this is their first formal academic presentation with an unfamiliar audience. For many teachers, this is the first time they presented to a group of peers.

A sample of the presentation schedule for one V-CCLS team can be found in Table 3.7. Progressing through the teams for a cohort is a significant investment in time.

V-CCLS presentation protocol (40 minutes)

Table 3.7 Example V-CCLS team presentation schedule

Approx. time	Step
20 Minutes	Hand out Form 1 (Presenters) Presenters present what they learned through V-CCLS • Observers take notes on Form 2.
3 Minutes	Silent writing • Observers individually reflect on what they learned silently. • Observers fill out Form 3. • Presenters can quietly discuss as a group how they thought their presentation went. No recording of this conversation.
4 Minutes	Warm feedback • Presenters fill out Form 4 during the warm and cool feedback. They do not respond to the feedback. • Observers rapidly give warm feedback to presenters. What went well? What was particularly useful? What was learned? • Presenters must remain silent and listen. • All observers would like a chance to provide their feedback, so please keep feedback very brief. • Observers should not repeat feedback that has already been presented.
4 Minutes	Cool feedback • Observers rapidly give cool feedback to presenters. What did not go particularly well? What was not useful? What opportunities were missed? • Presenters must remain silent and listen. • All observers would like a chance to provide their feedback, so please keep feedback very brief. • Observers should not repeat feedback that has already been presented.
3 Minutes	Personal reflection • All parties are silent. • Observers reflect on what they have learned. Fill out Form 5. • Presenters reflect on what they have learned. Fill out Form 6.
2 Minutes	Presenters' response • Presenters can now share with observers what they have learned from the presentation. • Observers must remain silent.
4 Minutes	Presentation preparation • The next V-CCLS group sets up their presentation and hands out their Form 1.

After presentations, the work of the V-CCLS groups is complete. All that remains is final documentation for the team (Table 3.8). Their work is captured in a digital or physical portfolio.

Preparing group portfolio V-CCLS

Table 3.8 Example of final documentation of V-CCLS teams

Due Jan 28, xxxx

Section 1: Please include a title page using the template that is posted in the Cohort 3 Google site.

Section 2: Please organize the set of forms from each lesson observed (Please find all required forms on the SEF website.)

☐ V-CCLS Form 1: Pre-Observation form
  ○ Form would have been filled out by teacher being observed.
☐ V-CCLS Form 3: Observation feedback form
  ○ Need one copy from each V-CCLS member who observed the lesson, including facilitator.
☐ V-CCLS Form 4: Demonstration Teacher Feedback Gathering Sheet
  ○ Form would have been filled out by teacher being observed.
☐ V-CCLS Form 5: Meeting Reflection Form
  ○ Need one copy from each V-CCLS member who observed the lesson, including facilitator.
☐ V-CCLS Form 6: Demonstration Teacher Reflection (may be after meeting).
☐ Copy of student work that went with the lesson.

Section 3: Please include all videos on a flash drive or OneDrive, or URLs, etc.
Section 4: Please include your presentation. If you have electronic files such as a PowerPoint, please create a folder titled "Presentation" on your flash drive, etc., and include these files.

Outcomes of the V-CCLS process

Teachers frequently cite the V-CCLS as the most powerful element of the SEF. Both elementary and secondary teachers express appreciation of the high level of content with which students are engaging at different levels. High school teachers admired the sophistication in some of the lessons conducted in elementary classes, noting that they were impressed with students' science discussions and use of vocabulary, for example. Elementary and middle school teachers felt that many of the science concepts in high school classrooms were quite complex, underscoring the importance of building students' conceptual foundations at a young age. Secondary teachers were struck by elementary teachers' use of reading strategies across the curriculum, as well as how these teachers seamlessly and constantly employed management strategies for materials, groupings, and discussions. Middle and high school teachers frequently remarked that they could benefit from some of the strategies they observed in teammates' classrooms.

Working across grade bands was incredible. As a high school teacher, there's this void of what their previous 8 years of science was like. I got to witness an incredible journey full of rich vocabulary (shoutout to the use of "schema" in elementary school!) and the joy of early discovery. In working with the teachers, we were able to understand more the developmental increments, pulling out key ideas for elementary level, expanding to middle, and expanding again to high school. It gave me a new way to talk about the science story.

Tal SebellShavit, High school, Cambridge, MA.

Watching lessons from grade 3 through AP biology in photosynthesis illustrated a very useful picture of how content moves through the grades and how we each teach the same content but how differently we do so. I never observed this before. (Survey response, NJ teacher, 2016)

Through research, we have uncovered several important themes that emerged from teacher reflections, interviews, journal entries, and other data sources. One important theme was that teachers developed a deeper understanding of how scientific knowledge develops from the early grades through high school (Gunning et al., 2020). Teachers had vague understandings of what students learned before, but watching their team members' videos from primary through high school classrooms allowed Fellows to better understand the progression of content and increasing complexity of ideas and practices in science. Using CTS grounded their observations and analyses in research and other shared lenses so that the teachers listened carefully to student responses and examined their work with an eye for understanding thinking.

Another important theme was the teachers' perceived professional growth that they attributed to the guidance and support from their vertical teams (Gunning et al., 2020). The type of professional growth they experienced varied widely, but regardless of years of teaching experience, which ranged from 3 to 30+, each teacher felt that their growth was significant. Some Fellows cited an increasing comfort level with inquiry-based learning, others with incorporating student collaboration protocols, still others cited improved ability to support emergent bi-lingual (i.e., multilingual learners) in their classes. Fellows felt that their growth was noteworthy and important, and that interacting with their peers at diverse grade levels impacted them in a positive way.

The second half of the year H-CCLS

The second half of the first year of the SEF program is still focused on instruction but changes from a vertical articulation of curriculum across grades to a horizontal application of a research-based teaching approach with an NGSS SEP within a grade band (K–2, 3–5, 6–8, 9–12). New teams are formed and will continue with cross-district collaboration. In this arrangement, Fellows will be much more familiar with the grade-level expectations and maturity of the students for which the SEP is the focus. In the H-CCLS groups, the manner in which the content is taught comes to the forefront.

Without exception, the Fellows are disappointed to leave their V-CCLS groups to form H-CCLS groups. We note this as a positive outcome of the depth of relationships and trust that formed during the V-CCLS cycle. We are grateful that new positive relationships form in the H-CCLS as well.

Over the course of the semester, the relationship between myself and my vertical team members grew. We were all excited to see how our teaching in the different levels (elementary, middle, and high school) connected. When the time came to move to our horizontal groups, we protested. We didn't want to leave the relationships we'd built with our teammates.

Regina Borriello, High school, Clifton, NJ

I enjoyed working with the teachers from the same grade band. We were able to collaborate, support each other, and share our professional growth. It allowed us to create a consistent and engaging learning experience for students, while also fostering a positive and cooperative working environment for teachers. We were able to leverage each other's strengths and experiences to enhance our instructional practices and contribute to a cohesive educational framework.

Raisha Allen, Middle school, Desoto ISD, TX.

The release of the Framework for K-12 Science Education and the derivative NGSS provided a new vision of science learning. Curriculum should reflect the three dimensions of DCIs, SEP, and Crosscutting Concepts. DCIs have always been at the heart of state and national standards. The SEP and Crosscutting Concepts are new additions that recognize and codify that the teaching and learning of science in the 21st century is more than learning a collection of facts. The V-CCLS in the first half of the year focused on DCIs. The H-CCLS process puts the SEPs center stage. The SEP as described in the NGSS are:

1   Asking questions (for science) and defining problems (for engineering).
2   Developing and using models.
3   Planning and carrying out investigations.
4   Analyzing and interpreting data.
5   Using mathematics and computational thinking.
6   Constructing explanations (for science) and designing solutions (for engineering).
7   Engaging in argument from evidence.
8   Obtaining, evaluating and communicating information.

One of the goals of the second semester of the SEF program is for teachers to take a deep dive into **one** of these practices. For this work, the Fellows will form grade band teams. Each team will select a course of study that includes one SEP and a research article that focuses on a pedagogical approach that can be adapted to their classroom lessons. Each H-CCLS group reads about the eight SEP and then discusses which practices they are most interested in pursuing. Using a list of "top 3 practices" from each group as a guide, the groups are each assigned one practice. Most teams get their first choice but assignments are made so that each team has a different practice. As in the V-CCLS work, the second part of the CoS is a pedagogical approach grounded in a research article that the team selects. The decision on the approach that will be the focus of the lessons requires the team members to peruse research journals and identify a research topic that interests them all, and has the potential to develop their group's SEP. They then read some of these research articles, discuss, and decide. Since the focus of the H-CCLS is to gain a greater understanding of one of the SEP, the instructional approach that the grade span group chooses should be related to helping their students successfully use these practices at their grade level. The research article the group decides upon should help teachers to incorporate specific strategies that foster the understanding and use of the specific scientific practice *by the students* such as "Helping Students Use Claims and Evidence," which is aligned to NGSS Scientific Practice #7: Engaging in Argument from Evidence. For example, a high school H-CCLS may dive deeply into "Planning and Carrying Out Investigations" – Practice #3, each in their own content specialty, while focusing on a research article to complete their CoS about "Using Scaffolding and Backward Design to help Students Plan Experiments." On the other hand, an elementary H-CCLS focusing on NGSS practice #7: "Engaging in Argument from Evidence," may chose a research article to learn classroom strategies that foster argumentation skills and help students understand what evidence is and isn't at the elementary level. More example CoS are provided in Tables 3.9 and 3.10.

Table 3.9 Example H-CCLS CoS

| Team grade span | Course of Study (CoS) | | |
	SEP	Title of research article	Research article citation
K–2	Integrated STEM: Planning and carrying out investigations	The effects of a STEM intervention on elementary students' science knowledge and skills.	Cotabish, A., Dailey, D., Robinson, A., & Hughes, G. (2013). The effects of STEM intervention on elementary students' science knowledge and skills. *School Science and Mathematics, 113*(5), 215–226. https://doi.org/10.1111/ssm.12023
3–5	Planning and carrying out investigations	Inquiry-based science education: Towards a pedagogical framework for primary school teachers.	van Uum, M. S. J., Verhoeff, R. P., & Peeters, M. (2016). Inquiry-based science education: Towards a pedagogical framework for primary school teachers. *International Journal of Science Education, 38*(3), 450–469. https://doi.org/10.1080/09500693.2016.1147660
6–8	Constructing explanations	Supporting students' construction of scientific explanations by fading scaffolds in instructional materials.	McNeill, K. L., Lizotte, D. J., Krajcik, J., & Marx, R. W. (200). Supporting students' construction of scientific explanations by fading scaffolds in instructional materials. *Journal of the Learning Sciences, 15*(2), 153–191. https://doi.org/10.1207/s15327809jls1502_1
9–12	Analyzing and interpreting data	Measuring graph comprehension, critique, and construction in science.	Lai, K., Cabrera, J., Vitale, J. M., Madhok, J., Tinker, R., & Linn, M. C. (2016). Measuring graph comprehension, critique, and construction in science. *Journal of Science Education & Technology, 25*, 665–681.

Table 3.10 A second example H-CCLS CoS

Team grade span	Course of Study (CoS)		
	Title of research article	Title of research article	Research article citation
K–2	Developing and using models	"How can I build a model if I don't know the answer to the question?": Developing student and teacher sky scientist ontologies through making.	Becker, S., & Jacobsen, M. (2019). "How can I build a model if I don't know the answer to the question?": Developing student and teacher sky scientist ontologies through making. *International Journal of Science and Mathematics Education, 17*(S1), 31–48. https://doi.org/10.1007/s10763-019-09953-8
3–5 Team	Analyzing and interpreting data	Students' successes and challenges applying data analysis and measurement skills in a fifth-grade integrated STEM unit.	Glancy, A.W., Moore, T. J., Guzey, S., & Smith, K.A. (2017). Students' successes and challenges applying data analysis and measurement skills in a fifth-grade integrated STEM unit. *Journal of Pre-College Engineering Education Research, 7*(1), Article 5. https://doi.org/10.7771/2157-9288.1159
6–8 Team	Developing and using models	Students' understanding of the role of scientific models in learning science	Treagust, D. F., Chittleborough, G., & Mamiala, T. L. (2002) Students' understanding of the role of scientific models in learning science. *International Journal of Science Education, 24*(4), 357–368. https://doi.org/10.1080/09500690110066485
9–12 Team	Engaging in argument from evidence	Facilitating argumentation in the laboratory: The challenges of claim change and justification by theory.	Walker, J. P., Van Duzor, A. G., & Lower, M. A. (2019). Facilitating argumentation in the laboratory: The challenges of claim change and justification by theory. *Journal of Chemical Education, 96*(3), 435–444.

The relationships among our H-CCLS team evolved and strengthened significantly throughout the semester. Initially, we came together with diverse backgrounds and perspectives, but as we collaborated on Claims, Evidence, and Reasoning, we forged strong bonds. Regular meetings and discussions provided opportunities to share ideas, experiences, and insights, fostering mutual respect and understanding. Over time, these interactions cultivated a supportive and collaborative environment where we felt comfortable exchanging feedback and supporting each other's growth as educators. Ultimately, the semester was characterized by camaraderie, trust, and a shared commitment to excellence in science education.

Candace Edmerson, High school, Grand Prarie, TX.

Our H-CCLS group chose the SEP practice of asking questions. I think it took us several attempts at understanding to truly get to the root of what the practice asks teachers and students to do and, at first, building lessons around student questions seemed like an impossible task. Years later, however, I can see that as a fundamental step towards a shift in my professional practice. In my classroom today, you'll find a driving question board filled with student questions that actively gets referenced and added to.

Kris Grymonpre, Middle school, Boston, MA.

Once the course of study (SEP and research article) has been agreed upon, the Fellows embark on the H-CCLS process, mirroring the V-CCLS process of the first semester. They establish an observation schedule where each teacher, in turn, will record a lesson, share a video of that lesson, and then the group debriefs that video using the tuning protocol that includes warm and cool feedback as shown in Table 3.11. These debrief sessions also include reflections on student artifacts from the lesson. Preparation for the debrief session requires each team member to view the video. During the debrief, team members are filling out the forms that will be saved and presented as part of their portfolio. Throughout their work together, they are aware that they are collecting artifacts and meeting recordings to submit after their conference presentation to the whole group.

Format of SEF H-CCLS presentations

Table 3.11 Protocol for H-CCLS

Prior to meeting

- **Presenters**
  - Prior to the presentation, presenters fill out "Form 1: H-CCLS Pre-Observation Form."
  - Presenters hand out completed "Form 1: Pre-Observation" to observing Fellows.
  - Presenters present 20-minute presentation on their H-CCLS work.
  - During warm and cool feedback, presenters take notes on "Form 4: Recording Form." This is to keep and include in their binders.
  - Once presentation and feedback are complete, presenters have three minutes to fill out "Form 6: Presenter Reflection Form." This is to keep and include in their binders.

- **Observers**
  - Read the pre-observation information (Form 1).
  - Listen to H-CCLS presentation.
  - During the presentation, observers take notes and write down any questions on "Form 2: Observation Notes."
  - After presentation, Observers have three minutes to fill out "Form 3: Feedback Form."
  - Observers provide warm and cool feedback to presenters (8 minutes total).
  - Personal reflection – Observers now have three minutes to reflect on what they have learned from this presentation. Observers fill out "Form 5: Observer Reflection Form." Observers will keep this and include it in their binders.
  - Observers hand "Form 3: Feedback Form" to presenters.

- **Facilitator**
  - Hands out forms to Observers (excluding Form 1).
  - Ensures all protocols are followed and are on time.
  - Brings a watch so that he or she can keep time during the meeting.

(Continued)

Table 3.11 (Continued)

- **Required forms**
  - ○ **Form 1, Pre-observation Form** – For presenters to hand out to Fellow observers prior to the presentation.
  - ○ **Form 2, Observation Notes** – For observers to take notes during the presentation and keep for themselves.
  - ○ **Form 3, Feedback Form** – For observers to write warm and cool feedback and give to presenters.
  - ○ **Form 4, Recording Form** – For presenters to take notes during warm and cool feedback.
  - ○ **Form 5, Observer Reflection Form** – For observers to reflect on what they learned from the presentation.
  - ○ **From 6, Presenter Reflection Form** – For presenters to reflect on what they learned from presenting.

The team will use the same protocol described above in the V-CCLS to debrief and provide feedback on each other's lessons and by doing so will gain a stronger understanding of how their chosen SEP is approached in the classroom. At the end of the semester, they will present their findings regarding their SEP and their research article's impact on lessons at their grade level. They will also hear from the other four teams. In this way, all Fellows will develop an in-depth understanding of **one** SEP and a familiarity with **four** other practices. They will also have a sense of the complexity of these practices and how they can be embedded in lessons. This intense exploration of a single SEP to classroom lessons contrasts sharply with the limited workshop approach that so many teachers experience through traditional professional learning workshops where all eight SEPs are introduced to teachers in a day or less.

Like the V-CCLS groups, H-CCLS groups will need to use time outside of the monthly SEF meeting time to discuss the chosen research paper, video lessons, watch observation videos, debrief the observation videos, look at student work, and prepare for a presentation. Groups will present findings and learnings from their work at the June SEF Teacher Leadership Conference.

Conference

The relationship among my team became more than just a necessary working relationship. After years, I continue to consider them friends and colleagues. From the start, we worked well together. We shared the load and supported each other. My partners came from different school districts and grade levels, and observing their teaching and sitting in their classrooms was enlightening. I lost track of how many times I saw one of them do something that I knew would help instruction in my classroom. None of us had experience presenting to large groups, and we were definitely nervous creating our presentation and presenting it to a room full of educators. However, we shared a common sense of humor and a commitment to each other. We laughed at and learned from our mistakes, and we did well when presenting to our Wipro peers. Later, when we presented to a full room of educators at CAST, I did not feel nervous at all. My team was there to back me up!

James Mining, Middle school, Irving ISD, TX

The presentation provided by this program is a great opportunity for us to reflect on our teaching experiences and gain metacognition about our growth. I received education in Asia, which was a teacher-centered model. The graduate education I received from the U.S.A and the teaching experience in California changed my teaching model from teacher-centered to student-centered. But it didn't fully change my mind yet. Through the presentation, I realized that this Wipro program help me to grow in many aspects. Especially it awakened my mind to find the core part of the student-centered model. I appreciate this program help me to change my mind from teacher-centered to student centered successfully.

Yichang Liu, High school, San Jose Unified School District, CA.

At the conclusion of the H-CCLS semester, teams prepare presentations that include an overview of the SEP and research article that comprised their course of study followed by summaries of what was learned by observing the lessons An example of instructions sent to Fellows in preparation for their H-CCLS presentation is presented in Table 3.12. The presentation may not have a set format but certainly has a time

Table 3.12 Example instructions sent to SEF Fellows in preparation for their H-CCLS presentations.

Purpose: Lead the other Fellows and participants so that they walk away with the same learnings/discoveries your H-CCLS group had during your H-CCLS cycle.
Format: The Spring Teacher Leadership Conference has several purposes:

- To present findings of H-CCLS teams.
- To learn more about the SEPs and how they can be implemented across the grades.
- To provide thoughtful feedback to H-CCLS presenters.
- To learn about what it means to lead from the classroom.

Your H-CCLS team's presentation should focus on what you have learned as a grade span team around the NGSS SEPs and your research article – i.e. your Course of Study. You can choose any format to accomplish this goal. You can do this through a straight presentation or a combination of modes. Some ideas that we have thought of, though you are not limited to these, are a PowerPoint, a panel, a skit, a fishbowl, an interactive activity, lead an inquiry-based learning activity, showing clips of video, etc. Be sure to carefully think about what you want others *to learn* and be strategic in how you communicate that to the group. Be creative and have fun!

After each presentation, everyone – including the other presenters –will have an opportunity to provide warm and cool feedback to the group so that everyone can learn and improve their own presentation skills. The Inner Circle will provide verbal feedback, while the Outer Circle will provide written feedback. This written feedback will be collected, transcribed, and shared with the groups after the conference.

Content: During the presentation, you may want to include some of the following information:

- Team name
- Course of study including:
  ○ Science or Engineering practice assigned to team.
  ○ The research topic and a description of research article findings.
- How the research influenced your lessons and debriefs.
- Annotated student work.
- How the research influenced the way you analyzed the student work.
- Challenges that you came across.

(Continued)

54   A Roadmap for Transformative Science Teacher Leadership

Table 3.12 (Continued)

- Surprises that you came across.
- Findings/lessons learned from engaging in this course of study.
- Findings/lessons learned from being involved in a Horizontal CCLS group.
- A document that outlines what your findings are/what lessons you learned and want us to walk away with (also should state your course of study and team name).
- What would you have liked to have done or gone into more with more time? Next steps your group would like to have taken if given more time.
- How the H-CLLS work reflects the goals of SEF.
  - ○ To create and support a corps of teachers and leaders.
  - ○ To institute a culture of active and reflective instruction.
  - ○ To improve teacher quality through vertical alignment within content and horizontal alignment within grade bands, meeting in small groups, and professional development in order to increase student achievement.
- Time for questions from other Fellows.

It might also

- Be entertaining.
- Be a way to share the resources you have found.
- Be a PowerPoint.
- Be interactive.

constraint of approximately 20 minutes. The presentation is expected to be an exemplary lesson that actively engages the participants and often includes parts of the video recorded lessons.

This conference, unlike the first semester V-CCLS presentations, usually includes invitees who are not part of the cohort. These can be Fellows from prior years or from other sites. It can also include principals and other administrators from the schools. The V-CCLS presentations, as noted, may be the first time that the Fellows have ever presented to peers. The H-CCLS presentations are often the first time that the Fellows present to other professionals with whom they have little or no prior relationship. The fear and nervousness experienced by some teachers in anticipation of their presentation is often replaced with a sense of confidence and accomplishment after the presentation. Fellows have remarked how they are much more ready to speak up and contribute at school faculty meetings because of their experiences presenting to Fellows in the fall and a larger audience in the spring. The presentation schedule is provided in Table 3.13.

H-CCLS presentation protocol (40 minutes)

Table 3.13 Example H-CCLS team presentation schedule

Approx. time	Step
20 Minutes	Hand out Form 1 (presenters) Presenters discuss what they learned through H-CCLS • Observers take notes on Form 2.

(Continued)

Table 3.13 (Continued)

Approx. time	Step
3 Minutes	**Silent writing** • Observers individually reflect on what they learned silently. • Observers fill out Form 3. • Presenters can quietly discuss as a group how they thought their presentation went. No recording of this conversation.
4 Minutes	**Warm feedback** • Presenters fill out Form 4 during the warm and cool feedback. They do not respond to the feedback. • Observers rapidly give warm feedback to presenters. What went well? What was particularly useful? What was learned? • Presenters must remain silent and listen. • All observers would like a chance to provide their feedback, so please keep feedback very brief. • Observers should not repeat feedback that has already been presented.
4 Minutes	**Cool feedback** • Observers rapidly give cool feedback to presenters. What did not go particularly well? What was not useful? What opportunities were missed? • Presenters must remain silent and listen. • All observers would like a chance to provide their feedback, so please keep feedback very brief. • Observers should not repeat feedback that has already been presented.
3 Minutes	**Personal reflection** • All parties are silent. • Observers reflect on what they have learned. Fill out Form 5. • Presenters reflect on what they have learned. Fill out Form 6.
2 Minutes	**Presenters' response** • Presenters can now share with observers what they have learned from the presentation. • Observers must remain silent.
4 Minutes	**Presentation preparation** • The next H-CCLS group sets up their presentation and hands out their Form 1.

The H-CCLS semester has provided an opportunity for each team to explore one SEP in depth. Each team member has observed numerous applications of one SEP in successive classrooms. This intensive focus on one practice over the course of a semester has allowed each Fellow to better understand the complexity and richness of the SEP. It has also modeled an approach that can be used in subsequent semesters on how to gain expertise in additional SEPs.

The conference has given each team a forum to teach other Fellows in other teams about their chosen SEP. Simultaneously, each Fellow has gained some familiarity with the SEPs chosen by the other teams. Assuming that the cohort of 20 Fellows had four teams, each Fellow hears presentations regarding three other SEPs. This approach is in sharp contrast to the workshops that teachers often attend where the expectation is that they learn about all eight SEPs in a few hours.

Summary

In this chapter, we have presented the heart of the first year of the SEF program. You have read quotes from Fellows as they reflected on their experiences. The CCLS (vertical and horizontal) approach can be very powerful. Some may choose to simply implement these aspects of the program or even just one type of CCLS, depending on the goals of the school district. The protocols are flexible yet robust enough to stand alone. They are even more powerful when combined. The protocol should not be tweaked. The protocols create and then protect the culture of trust and safety that are necessary to have the open, honest conversations in the group. The program achieves its ultimate goal by combining this first year work with the individual leadership work of the second year, discussed in the next chapter. Before we close this chapter, we provide more testimonials and reflections from past Fellows.

In my quest for collaborative opportunities, I found it in the V-CCLS cohort. In our video lessons, I witnessed elementary students making scientific observations and explaining their reasoning. The sense of wonder and excitement was evident, and I learned so much from my vertical team. We transitioned to our H-CCLS to focus on graphing and interpreting data skills. The team of Freshman Physics teachers from 4 different districts working together was refreshing. By partnering with teachers from different districts, it re-energized my thinking around intentional instructional practices we developed tools I continue to use in my classroom.

Marsha Tyson, High school, Columbia Public School District, MO.

I have enjoyed working with teachers from different grade bands for many reasons. The collaboration had excellent benefits not only for my teaching but for the school climate, and the relationships amongst all educators involved. Horizontal and vertical group collaborations afforded us a sounding board for new techniques, discussions around strategies that worked, and suggestions for things that might not have worked so well. We were able to self-reflect on professional development, curriculum design, and ways to integrate STEAM into weekly lesson plans. The opportunity provided to me through Wipro was the perfect setting for shared instructional strategies. This year, I was able to work with teachers from two other schools in my district to plan their first Family Steam Night. It was a huge success. The event greatly impacted the community and brought recognition to the school and to the idea that STEAM is an integral part of their children's everyday learning.

An Marie Manganiello, Lower elementary (K-2), New Rochelle, NY.

When I first started [SEF], I did not realize how this program would impact my personal and teaching career for the better. [SEF] pushes teachers to get out of their comfort zone and to embrace change in order to improve their classrooms. Personally, it takes time for me to get comfortable and come out of my shell around individuals who I do not know well. However, after participating in [SEF], my confidence level has definitely increased. In addition, throughout these past two semesters I have enjoyed sharing ideas with my fellow teachers and supporting one another as we work together in order to achieve a common goal. As a teacher, you get to a point where you get used to your daily routine and become very comfortable in a particular type of teaching style. However, [SEF] has given teachers the opportunity to communicate with a variety of other teachers in surrounding school districts which helps with discussing the positives and negatives of our lessons. As a result of this discourse, I am able to take others feedback and incorporate their critiques and ideas into different areas that will greatly benefit my classroom. (Lisa Godina, Middle School, TX)

Notes

1   We use the term semester throughout to recognize that the first year of the program is split into two equally important portions. While many colleges and schools follow a semester schedule, we recognize that all do not. The SEF program need not match exactly with a school calendar – rather the V-CCLS and subsequent H-CCLS (discussed later) should be given approximately equal time during the school year.
2   DuFour, R. (2004). What is a "Professional Learning Community?" *Educational Leadership, 61*(8), 6.

References

Allen, D., & McDonald, J. (2003). The Tuning Protocol: A process for reflection on teacher and student work. https://www.clee.org/resources/tuning-protocol/

Gunning, A. M., Marrero, M. E., Hillman, P. C., & Brandon, L. T. (2020). How K-12 teachers of science experience a vertically articulated professional learning community. *Journal of Science Teacher Education, 31*(6), 705–718.

Hord, S. (1997). Professional learning communities: Communities of continuous inquiry and improvement. (ED Publication No. 410-659) Office of Educational Research and Improvement (ED), Washington, DC.

Hord, S., & Sommers, W. (2008). *Leading professional learning communities: Voices from research and practice.* Corwin Press.

Keeley, P. (Ed.). (2005). *Science curriculum topic study: Bridging the gap between standards and practice.* Corwin Press.

Mundry, S., Keeley, P., & Landel, C. (2009). *A leader's guide to science curriculum topic study.* Corwin Press.

Chapter 4

The Growth Plan System (GPS) year

Arthur Eisenkraft

The great thing about my GPS [Growth Plan System] is that it will live on beyond [Wipro]. I am scheduled to present at the MAST (Massachusetts Association of Science Teachers) conference, will put in an application for NSTA (National Science Teachers Association) and will continue to work with the Boston Student Advisory Council (BSAC) students around promoting and engaging others into the climate change curriculum project. Working on the GPS has allowed me the opportunity to find an avenue to engage in a project that is something I care deeply about and enjoy working on. I have always been involved in activities and projects outside the classroom, whether it was summer or afterschool programs, but this is the first project I have complete control over and have leveraged this product into some successful opportunities. (Tim Gay, High school, Boston, MA)

The quote above from this Science Education Fellowship (SEF) Fellow highlights the value of the second year of the Wipro SEF journey. The nature of the Fellows' work shifts from teachers diving into their instruction in teams to individual teachers exploring teaching issues of importance to them and practicing their leadership skills; this is a significant step for Fellows. Until this point, they tend to be comfortable in peer groups where they are seen as teachers reflecting on their classroom practice, and are willing to share their knowledge with their grade band (horizontal team) or discipline group (vertical team). This step asks Fellows to push themselves beyond their comfort zone and lead a school or district-level initiative and professional development. This is a new endeavor for most teachers, and while it is an exciting experience, it can also be a bit overwhelming.

In addition to being transformational professional learning for the Fellows, the SEF has the larger goal of a district transformation program focusing on science education and teacher leadership. As discussed in the previous chapter, during Year 1 of the program, participating classroom teachers (i.e., Fellows) engage in two discrete collaborative coaching and learning in science (CCLS) semester-long reflections on teaching. The first iteration is a vertical CCLS where a team of Fellows explore the vertical

DOI: 10.4324/9781003490586-5

articulation of curriculum from kindergarten to high school. The second iteration is a horizontal CCLS where Fellows regroup within grade levels and focus on one science and engineering practice. They also attend cohort meetings led by the IHE as well as district meetings led by their district science coordinator. The first year of the program ends with a conference where the CCLS teams present their findings.

Year 1 tasks are well articulated, defined, and have a common structure for all Fellows. The format and frequency of CCLS meetings are determined by the program. Fellows participate in cohort and district meetings where the agenda is set and facilitated by others to support them as they gain insight and develop important understandings. In this way, Year 1 is similar to enrolling in graduate-level courses; all teachers are quite familiar and comfortable with taking another university course. There are certainly elements of Year 1, such as sharing their classroom teaching through video recordings and presenting to a larger group of their peers, which may be daunting, but Fellows support one another and these anxieties dissipate quickly.

Year 2 of the SEF program is quite a contrast from Year 1. In Year 2, each Fellow creates, implements, and leads a professional development plan for the year. This Growth Plan System (GPS) requires each Fellow to take a deep dive into teacher leadership and sets the stage for their continued journey as teacher-leaders. These projects represent the diverse interests of the Fellows as well as the focus of their district's goals. Despite the challenges of this work, or because of them, Fellows take great pride and satisfaction with their efforts. Past GPS projects have included:

- A high school teacher started a school beekeeping project and enhanced place-based field trip options for others.
- A middle school teacher organized their colleagues across the district to align their laboratory experiences across buildings, which had never been done before. This allowed students to have similar experiences in key lab skills and activities during their middle school years.
- A high school teacher took it upon herself to organize and distribute physics take-home kits for 600 students.
- An elementary teacher planned and conducted virtual science nights for students and parents that encouraged science learning at homes supported by parents who learned how to frame questions around daily home activities.
- A middle school teacher wrote and illustrated a book about an immigrant middle school student and her science fair experience. Subsequently, that book was read by preservice teachers as part of their coursework.

Year 2 begins

In the final monthly meetings of Year 1, discussion with the Fellows begins about planning their individual GPS that they will implement in the coming school year. In these initial stages, Fellows are given the framework of the GPS and asked to think about how they can support district initiatives in their school and district while also defining personally important work that is tied to improving science teaching and learning. They are also encouraged to use the program to support them in initiatives they might not otherwise be able to pursue. In the final Year 1 monthly meeting and/or at a Teacher

Table 4.1 A SMART goal template for Fellows for their GPS

Developing SMART goals for your GPS Project:

Provide some background around why you are setting this goal.
What is your interest and motivation?
What is your vision of what you will accomplish?
How is this goal related to your personal and district goals?

S – Specific and Strategic: Who and what does this goal support? (you, your school, your district, your school's science department, your students, the broader STEM community...)
How does it support each of the identified entities?
How is this an individual and district-aligned goal?

M – Measurable: How will you measure your progression toward meeting this goal? What will be your benchmarks?

A – Actions: What are the actions you will take to meet the goal? (These actions will be part of your plan-consider them your "To Do" list. What do you need to do this? Do you need to purchase materials, buy books, attend a conference, take a class, etc.?

R – Rigorous, Realistic, Results: What rigorous, realistic results-focused evidence will be provided to show that you are working toward your goal? How will you know if you need to make mid-course corrections?
What rigorous, realistic results-focused evidence will be provided to show that you have met your goal? How might this goal help you to increase student achievement?

T – Timeframe: In what timeframe will key actions be completed and benchmarks achieved? How will you track your progress?

Leadership Conference, Year 1 Fellows are asked to articulate and share their preliminary GPS ideas through conversations with their colleagues, their District Science Coordinators (DSCs), and the SEF leadership team so they can receive feedback and ideas to help their plan take shape. Fellows are given the summer between Year 1 and Year 2 to develop their ideas into a cohesive plan to deepen their own content knowledge, professional skills, and capacity to lead others. They will accomplish this by setting goals, meeting regularly with a SEF GPS Advisor, their DSC and other Fellows, and establish outcomes and benchmarks toward their goals.

Fellows put their plan in writing. They are asked to set at least two professional goals, one related to and aligned with district initiatives (their District-Related Goal), and one that engages them in something of greater personal interest, yet still related to their science classroom teaching (their Personal [Professional Learning] Goal). Fellows are asked to articulate these goals as SMART goals: Strategic, Measurable, Action-oriented, Rigorous and Results-oriented, and Time-tracked (Doran, 1981). A template for these goals is provided as Table 4.1.

For each goal in their GPS, Fellows are asked to do the following:

- Be specific when describing their goal.
- Describe how their goal is both district-related or personally motivated.
- Be strategic by ensuring the District Related Goal serves an important purpose of the school or district as a whole and addresses something that is likely to have an impact or addresses student achievement in their classroom.

- Be ambitious and write goals that require focus and effort, but not be out of reach or overly ambitious.
- Make sure the goal is measurable in terms of quantity, quality, and/or impact.
- Provide some background and context for their goal.

In planning their work, they are asked to do the following:

- Envision the results by imagining what they will have accomplished by the end of Year 2 if everything were to go as planned.
- Set specific benchmarks to measure the process of the work and the progress of the outcomes.
- Describe clearly the key actions and steps that will be taken to achieve the goal.
- List the support (financial and other) they will need to complete their goal and align this support to their key actions.

Fellows are asked to think about the time they spend on their GPS in this way: 125 out-of-school hours should be devoted to SEF over the school year. Approximately 25 of the Fellows' hours in Year 2 are for cohort meetings, meetings with advisor, and meetings with their DSC and other district Fellows, which leaves about 100 hours for their GPS.

Fellows may divide these remaining 100 hours according to the needs of their individual personal and district goals: GPS hours should be thought of as two buckets of time, which can each hold a minimum of 30% and a maximum of 70% of their time. One bucket is the personal (professional learning) goal for the Fellow. This is time that the Fellow can set aside to do something of their choice to grow professionally; the other bucket is the work in support of district initiatives that they select through conversations with their district coordinator and principal. As an example, the Fellow may determine that their personal plan will take 45% of their time (45 hours) while the district-oriented plan will take 55% of her time (55 hours). Alternatively, another Fellow may split their time between personal and district-aligned plans at 35% and 65%. The advisor and the Fellow's DSC can help the Fellow work through realistic time expectations. A template for Fellows to plan their GPS is included in Appendix 4A.

Some Fellows will have a difficult time choosing their project and goals. Others will have a difficult time narrowing from dozens of ideas in their heads. The DSC and the advisor from the local university can help with this. Here, we provide some example goals at different grade levels and an example timeline. The dates in the timeline are flexible and are given here to show the number of steps that must be accomplished and to offer a suggestion for when different pieces should be accomplished.

Table 4.2 shows some sample GPS ideas and how they align with individual and district goals. Some samples from various grade bands are also included.

Year 2 of the SEF program is a year in the professional life of a teacher where they can choose to pursue something that they deem important. Many teachers have projects that they would like to pursue because they think them worthwhile and suspect that they will be beneficial to their teaching and their students. Time slips by, another year passes, and they don't find the time to engage with that project as other demands are placed on their day. The SEF program not only allows them to pursue these projects but holds them accountable for them. The SEF program encourages Fellows to do

Table 4.2 Brief illustrations of personal and district-aligned goals

Grade	Individual goal	District goal
K	To create a useable technology library for integration of science and technology in Kindergarten and Grade One. This will be a document created throughout the school year to provide staff with options in integrating technology with our new science curriculum.	To develop and implement a life science curriculum and usable space to aid in early childhood social and emotional growth via the science curriculum throughout the next school year.
2	To research teacher stress and the causes of teacher burnout. I will then create a handbook that includes indicators of teacher stress, plans for maintaining continued wellness, and strategies for dealing with acute attacks of stress in the moment.	I will create lab activities and supporting materials aligned with Next Generation Science Standards (NGSS) Practice #3: Planning and Carrying Out Investigations.
MS	To foster scientific self-guided journeys through the art of questioning the world around you.	With the incoming standards, several changes will need to be made to our current district curriculum. Together with my other district cohort, we will work on creating a cohesive unit to teach one of the strands of the new 8th grade standards.
HS	To use research to guide the creation of homework and track completion and content retention (as measured on quizzes). Since we are on a semester system, the first semester will be research and preliminary implementation. The second semester will be edits based on data in the fall semester, as well as gather more data to see if results are consistent with a new group of students.	To use the Science and Engineering Practice (SEP) to increase scientific analysis skills in students with lower academic fluency, and allow "stretches" for the more advanced students. Wants to focus on Analyzing and Interpreting data, Constructing Explanations, and Engaging in argument from evidence.
HS	To use cartooning to create a 180Dayz teacher comic graphic novel that will help new teachers avoid common pitfalls, obtain necessary skills, knowledge, etc., to become a successful teacher and improve student learning within their first year of teaching.	To realign current BPS Physics curriculum with new state science standards (what resources can we use, what needs to be shifted across grades or subjects, is there new content required that we do not have the materials for teaching and need to obtain, etc.)

something that addresses their own interests and something that speaks to their core – the personal goal. Through their GPS, they are defining and guiding the initiatives that they think will be beneficial, rather than following the paths that others have mapped out. Some Fellows really run with this freedom and embark on projects that are eclectic while others stay closer to more traditional professional learning projects.

Their second goal – the district-related goal – provides the Fellow with the opportunity to create the path toward that district goal. As they pursue their plan, they have a turn at "steering the school ship" as well as developing leadership capacity. They work with other teachers and with the encouragement of their DSC and their principal add another dimension to the traditional ways in which their school moves forward with initiatives. They may even transform the way in which teachers get involved and/or how school faculty meetings are conducted. Meeting templates and benchmarks are provided to Fellows with a level of detail that supports them and keeps them moving forward throughout the year. These are for guidance and each Fellow reviews these with their mentor at the university and their district science coordinator. Two examples are provided in Appendix 4B. In brief, the Fellows will:

- Choose, draft, and finalize GPS project description and goals with the consultation of their DSC, mentor, Institution of Higher Education (IHE), and school principal.
- Begin their portfolio which will be a record of the year's work.
- Embark on their GPS project (personal and district-related) goals.
- Attend meetings with their mentor, IHE, and their DSC.
- Engage with other teachers in their school on aspects of their work.
- Complete their portfolio and summarize their work in a GPS poster that will be shared with other Fellows, other teachers, and, perhaps, principals or other district personnel.

How does Year 1 prepare the Fellows for Year 2 (GPS)?

Different elements of Year 1 work help prepare the Fellows for Year 2 and support the second and third pillars of the SEF program – adult learning and teacher leadership. Some of these elements are implicit (e.g., CCLS protocols) while others are explicit (e.g., readings of research articles on teacher leadership, viewing of GPS posters from past Fellows).

The bulk of Year 1 is devoted to the CCLS meetings where four or five Fellows develop a course of study and share recordings of their classroom lessons and student artifacts. There is no one leader. They take turns being the leader, and so, in essence, they are all leaders. This realization that leadership is shared and that this model works so effectively helps set the stage for their Year 2 GPS. The Fellows are asked to keep this model in mind as they move forward in working with other teachers in their school or district. They are warned that if they begin their work by declaring that they are the leader, they are certainly headed for a more difficult path than if they explain that they will help frame the work that the teachers will embark on and that everyone will have an opportunity to share, contribute, and lead.

During Year 1, some of the monthly cohort meetings, which are facilitated by the IHE, include readings and reflections dealing with aspects of teacher leadership in addition to handling logistics of the program and supporting Year 1 CCLS endeavors. These discussions help Fellows learn about the landscape of research studies focusing on teacher leadership including the reviews by York-Barr and Duke (2004), Taylor et al. (2011), and Wenner and Campbell (2017). Some of the literature requires the Fellows to consider how to be teacher-leaders without leaving the classroom and without having the title associated with leadership (e.g., coordinator, chair, specialist).

The quarterly meetings of the Fellows in each district are facilitated by the district science coordinators. These meetings include the Fellows in Year 1 and, as time passes, adds the Fellows in Year 2 of the program. Some of the time at these meetings is allocated toward how the program is meeting the needs of the teachers and how the interactions of Fellows from one district are impacting the Fellows from another district. The most important aspects of these district meetings are for the Fellows to become acquainted with their own district priorities and to build a positive working relationship with one another. Too often, most teachers hear about district priorities after decisions have been made and many high school teachers are unaware of decisions concerning the elementary schools within their district. For example, the adoption of a new science program in grades K–5 rarely concerns high school teachers. The impact of that new elementary science program on high school content and approaches has usually been considered by the district administration including the DSC, but often has not included high school teachers. SEF, as a district transformation program, seeks to build a better, stronger, more inclusive district leadership where these K–12 science teachers can be assets to the DSC and, in turn, to the principals of their schools. The quarterly meetings provide the first opportunities for the DSC to share district wide ideas and initiatives at the early stages and benefit from the teacher perspective at different grade levels and across different schools. Ideally, the DSCs will learn to seek ideas, direction, and guidance from the team of Fellows, so that they have teacher input before planning district direction. The Fellows become a DSC "cabinet" of teachers who work together, having built common knowledge and trust from their shared experiences, and who have earned each other's respect.

Another activity from Year 1 that prepares the Fellows for their GPS is the viewing of the GPS posters from Fellows completing Year 2. (The first cohort of teachers will be the first to create posters. In that year, posters from other sites or from the SEF archive are shared with that group of Fellows.) One of the deliverables for all Fellows at the end of their GPS year is to create a poster summarizing one aspect of their GPS. These displayed posters provide a landscape of projects of the past and serve as a stimulus for the Fellows who are thinking about how they can mold their interests into a suitable project. At a poster presentation meeting, the Cohort 1 Fellows stand beside and present their posters and the Cohort 2 Fellows can ask questions of someone who has just finished the journey they will be embarking on. Along with the monthly meetings of the cohort and the quarterly meetings of the district, the poster session is intended to create excitement among the Fellows, and not intimidation by the work ahead.

During the culminating event of Year 1, Fellows share their initial ideas for their GPS through a "speed dating" activity. In the activity, each member of one-half of the cohort has three minutes with a Fellow, DSC, or person from the IHE to describe the project they are considering and receive feedback. After three minutes, the Fellows rotate and repeat their three-minute elevator speech and get additional feedback. This continues until the Fellow has heard from at least ten others. During these ten speed dates, the Fellow can adjust their presentation as they see fit, often including insights from earlier comments. The process is then repeated for the half of Fellows who have been offering feedback to now individually share their ideas and goals. Outcomes of speed dating include a better sense and refinement of the initial GPS project idea, a better articulation of the goals, and many helpful suggestions as to how to augment and improve the project focus.

How to support Fellows during Year 2 (the GPS year)

The sense of community that develops among the Fellows, the DSC, and the people from the IHE is a significant feature of the SEF that has enabled its success. Embarking on an independent project of the magnitude of a GPS does not require a Fellow to navigate alone. Fellows need a safe space to be successful and that requires emotional support, intellectual support, and logistical support. In this regard, the Fellow may just need someone who will listen or may also need someone who can help them reappraise their goals. In fact, the evaluations have revealed that the meetings with other Fellows and their relationship with their GPS mentor are of prime importance to the Fellows in Year 2 of the project. A Fellow's goals may be too ambitious and may need support and encouragement to limit, modify, or even give up one of the goals. Project leaders have had Fellows who were so overwhelmed with their original project and the sense that the hurdles were too high, that they thought the only solution was to quit the program. Listening, discussion, empathy and a reappraisal of the goals provided enough slack that they realized that they could be successful with a more limited approach.

Some Fellows require someone who can provide intellectual support. The Fellow may have a global sense of the goals and is not able to articulate the details of all that must be done for each part of each goal. Alternatively, they may be a detail person who loses sight of the big picture of what should be accomplished. Support can be simply assistance in balancing the large goals and the details and using project management tools that are available. Support can be reminders of how the goals fit into the district goals with a clarification of what those district goals are and how they are being executed. Often, the IHE mentors provide support in the sense of encouragement.

Other Fellows may need logistical support. They may be unaware of materials, books, supplies, and other resources available to them at the school, in the district, or that can be borrowed from the university or a local outlet. There may be external funding opportunities for their project. An elementary school Fellow may not realize that the high school has lab equipment that would be perfect for them to use in their work.

Emotional, intellectual, and logistical support can be provided by other Fellows (past or present), the DSC, the IHE leadership or a mentor assigned to the Fellow. Each person in this support network can bring different perspectives and guidance. This depth and breadth of experience and expertise is a defining feature of the SEF community. Table 4.3 gives a few, but hardly exhaustive, suggestions of how the community can support each other.

This support takes place in the planning of the GPS and during the GPS year. The network of support serves as both critical friends and shoulders to lean on. Some of this support takes place in smaller, private meetings while other support takes place in the meetings with the entire cohort. As Fellows struggling with their own work, the requirement to support other Fellows provides perspective, empathy, and a sense of expertise. Helping others with their challenges in completing their projects provides another opportunity for Fellows to learn to be a leader. Any teacher who has ever had to teach a concept that was initially at the edge of (or beyond) their comfort zone can relate to how being asked to explain or teach something helps one understand the material more deeply. A summary of the meetings during the GPS year illustrates the levels of support (see Table 4.4).

Table 4.3 Support roles in the GPS year

	Emotional	Intellectual	Logistical
Fellows (past/ present)	Empathizes; relates their own experience	Shares interest in their project and how it relates to their project	The help that they received while they were completing their GPS
DSC	Checking in; reminding the principal of the work being done	Clarification of district goals	Resources available at the school/ district level
IHE	Reassurance that projects of this sort are difficult	How others have approached projects like this; what the research literature may say about the initiative	Locating research articles and other resources through the university library
Mentor (may be part of the IHE team or community members with a particular expertise, i.e., informal learning staff, business partners, etc.)	Provide the reassurance of a critical friend who is not in an evaluation role	Shares interest in the goals of the project	Guidance on timetable; progress charts

Table 4.4 GPS year meetings

Advisor/mentor meetings	Advisors and assigned Year 2 Fellows	Each Fellow is assigned an advisor/mentor and will meet with this person for approximately eight hours during Year 2 of Cohort Fellowship to discuss progress, successes and challenges of their GPS work. Some advisors hold some of these meetings with multiple advisees and find that the conversations are richer as they share information across their GPS projects.
Callback meetings	All Year 2 Fellows, District Science Coordinators, SEF staff	In Year 2, the Cohort meets at least four times a year to discuss progress, successes, and challenges of their GPS work.
Quarterly district cohort meetings	District Science Coordinator and all cohorts of Fellows from one district	Four to six times per year, in both years of cohort fellowship and afterward. All the Fellows from all cohorts from one district meet on a regular basis with their District Science Coordinator. These meetings promote a leadership community among Fellows. The DSC can share district priorities and plans with the Fellows as they unfold. When the Fellows are in Year 2 of the program and working on their GPS, this is a good forum for them to share their district-wide goals and their work and to get feedback and support and guidance from the other Fellows and from their DSC.

Meeting with SEF GPS advisor/mentor

In Year 2, the Fellows implement their GPS while meeting regularly with their assigned GPS advisor. They are required to meet at least once a month with their advisor, either face-to-face or through tools such as Zoom, etc., so the advisor can provide support, check on the progress of the GPS plan, and help them revise or refine as needed. The GPS project goals are not set in stone. As with any creative and boundary-pushing initiative, unforeseen issues arise. These may be opportunities or insurmountable hurdles. Through discussions, the Fellow may come to the conclusion that the goals were too ambitious and changes are required. The mentor can provide the support needed to move forward or to back off on specific elements from the original GPS plan. It often happens that the original plan is too ambitious, and the Fellow becomes disappointed and frustrated. The advisor can "talk the Fellow down" and help them to reorient their expectations. The GPS advisor is there to help the Fellow over stumbling blocks or to provide the Fellow with fresh ideas. The idea is that the advisor is someone with whom the Fellow can make a connection and who might have a particular strength in the areas concerning the Fellow's plan, but does not play a district supervisory position to the Fellow.

Year 2 callback meetings

The Year 2 Fellows meet at least four times in the second year of the program (in contrast to the monthly meetings of Year 1). They share their success and challenges implementing their GPS by this point in the year. The cohort of Fellows had worked together the past year in both V-CCLS and H-CCLS groups and they developed trust and recognize the value of giving warm/cool feedback to each other. There are multiple strategies for getting the Fellows to discuss in small and whole groups different aspects of their work at different parts of the year. Assisting each other helps solidify the learning community of the past year.

Year 2 district cohort meetings

Each year of the program, a new set of Fellows are identified from each district. In the first year of the program, the four Fellows from one district meet with their DSC. These meetings are concerned with the larger SEF program as well as district interests and challenges. The DSC usually has a more general sense of science offerings across the district as well as possible interests and initiatives of the superintendent and other district administrators. These meetings bring the Fellows into the conversation of district issues from which many school and classroom policies emerge. In the second year of the program, an additional four Fellows –Cohort 2 – from the district will now join the four Fellows in their GPS year from Cohort 1. The quarterly district meetings now include eight Fellows and the district science coordinator. In this second year, discussions of similar topics from the first year continue. In addition, the Cohort 1 Fellows, now engaged in their GPS, can share their goals and their progress. In the third year of the program, the final four Fellows join the district science teachers where the 12 Fellows and District Science Coordinator can share the general district concerns and hear about the new GPS projects. Cohort 1 has completed the formal part of the SEF (Year 1 and Year 2), and as emerging leaders, are encouraged to attend the meetings with the DSCs and continuing cohorts of Fellows and expand their role as teacher-leaders, offering guidance and support.

My GPS project was integrating my ELA and Science instruction. When I got started, I remember feeling really overwhelmed and not knowing where to start. Having support from different members of my Wipro cohort and advisors helped me to shape the project and find direction. My students were so engaged in the work we were doing and now I continue to use this framework as I do the science and math instruction in a co-teaching format. Having the framework helped me to support and collaborate with my co-teacher in her ELA instruction as I was her mentor this past year.

Josie Hess, Upper elementary (3-5), Community R-VI Elementary, MO.

Throughout my GPS project, I felt as if I had an auditorium full of peers who were cheering me on and wanting me to succeed. Their encouragement helped me to step outside of my comfort zone and approach administrators throughout my entire district to create STEM Family Nights.

Monique Dewar-Dituri, High school, Clifton High School, NJ.

Communication

Communicating plans, progress, and conclusions are crucial elements in the success of all projects. The first set of communications by the Fellows focuses on the GPS ideas as described earlier. Using knowledge of district priorities and sharing their interests, each Fellow converges on a GPS plan that is approved by the IHE and the DSC. The principal of the school is often brought into the conversation as well. In our experience, gaining the principals' support is key to the sustainability of the projects beyond the GPS year. Awareness of the GPS project is a low bar for any principal; effective principals display interest in the GPS project and provide encouragement. Highly effective principals would actually weave that teacher's work into the fabric of the school in some way.

During the GPS year, each Fellow keeps a journal and provides updates on progress to the DSC, the IHE, and the mentor. During callback meetings of the entire cohort, each Fellow also shares their progress with other Fellows. Progress reports provide accountability, periodic encouragement, and a scheduled time where Fellows can express their frustrations as well as their successes. Some of the best interactions of Fellows at callback meetings have been with sharing of their respective journals (including their weekly reports of "Wows" and "Yikes"). An example set of journal entries and reflection from one Fellow is provided in Table 4.5.

Each GPS project includes an obligation to involve other teachers in the project or in a professional development session reflecting on the project. Enlisting the teachers, planning a session and evaluating the session is another facet of communication during the GPS year. The project also situates the Fellow as a teacher-leader in their school and district (Taylor et al., 2022).

Table 4.5 Sample journal entries and reflection from one Fellow

Friday, October 28	
What I have done	
	Poster, graphic organizer, and two labs involving writing procedures. Research on teacher stress and methods for dealing with teacher stress
Wows	
	The amount of information in the world about both of my topics is pretty high. It is great to see what else people are doing with writing procedures and remedies for teacher stress.
	Although the information on writing procedures is out there, materials for helping students write procedures was not readily available (graphic organizers, sentence starters, etc.) so I'm glad I have this opportunity to put this stuff together.
Yikes	
	I am having such a hard time keeping up with all of my commitments, including [SEF], mentoring a student teacher, having 155 students, being a class advisor, and general life.
	I keep reading that taking care of my health is important to maintaining a lower level of stress but I feel like I don't have time to actually do that. I made a lot of progress with my personal health this summer but I am backsliding.
Sunday, December 18	
Wows	
I have somehow managed to handle being given 160 new students the day after Thanksgiving weekend.	
Yikes	
	I was given 160 new students the day after Thanksgiving weekend.
	Many of these students didn't even have notebooks.
	I have 4 months of chemistry to fit into 6 weeks to prepare these students for the midterm in January.
	Everything else, including this fellowship and my own life, has completely fallen apart because 100% of my time is dedicated to my job and my new students now.
Friday, January 20	
Wows	
	Midterms are over so I can return to a normal pace.
	I did manage to make some progress on my GPS.
Yikes	
	I'm sort of lost in all of my projects. I'm not sure where I am or where to go.
	I feel so behind in [SEF].
	I also have a big show to run for the school that is starting to ramp up. I had planned my GPS so that I had less to do in February and March but now I need to do more to catch up. It is all very overwhelming.

(Continued)

A Roadmap for Transformative Science Teacher Leadership

Table 4.5 (Continued)

Sunday February 5	
What I have done	
	Three labs, reflection One PD Started my 3rd teacher stress project
Wows	
I am getting caught up – I'm about halfway done with both parts of my GPS.	
Yikes	
	I am still incredibly busy and incredibly stressed I reached out for help and/or guidance and didn't even receive an email back. That was months ago now. I'm not sure where to go or what to do.
Sunday March 19	
Wows	
	I am actually completely caught up and back on track. I am really looking forward to a calmer fourth quarter after an insane year. I am proud of myself for doing this completely on my own. I didn't get a lot of support from outside sources.
Yikes	
	I had a major project going on at school so I had to put everything else out of my mind. I am using this weekend to get caught up on things, including this fellowship. I also have to work on the poster and prepare for the poster session. This is incredibly hard to balance with the year I have had.
Monday April 17	
Wows	
Looking forward to this being done!	
Yikes	
	Spending the first day of my vacation working on this poster. I still feel very overwhelmed, especially when I think about putting the portfolio together this summer. I haven't had the time this year to be as organized as I usually am.
Sunday April 30	
Wows	
	It feels really good to be back on track. No more than 10 hours each week is manageable for this project, and that is what I'm doing now.

(Continued)

Table 4.5 (Continued)

Yikes	
	Some of the students, especially the CP students, are really struggling with writing their own lab procedures. I am struggling to find teachers to help with my personal GPS. I am very nervous about the poster session! This has been such a difficult year and I'm not looking forward to putting it on display.
Monday June 5 Wows	
	The poster session was an enjoyable experience and it was really nice to see what else people had been working on I think I should pick a personal goal every year and make it something more enjoyable for me (see yikes below)
Yikes	
	I can't help but wish this year had gone differently for me. I didn't get as much out of this second year as I could have if things had been more regulated with my school workload. I also regret not picking a personal goal that was more enjoyable. I saw that someone had chosen to read books for pleasure as their goal. Something like that would have been more beneficial to reducing my stress level than the project I chose about studying methods for managing stress levels.
A Sample Reflection from this Fellow	
	Overall, I learned a lot about my strengths and limitations as a teacher this year. I learned that sometimes I take on too much and that directly impacts my teaching and my stress level. But I also learned that I can be successful even when the odds are stacked against me. As I read through my journal, I could feel the stress I felt all year again. However, towards the end of the year I could also feel the pride and the sense of accomplishment. I am happy with the strides I made this year and the lessons I learned. I hope to put all these lessons into practice in future years and create more of a balance between excelling in the classroom and creating a less stressful environment for myself.

At the conclusion of the GPS project, communicating the results of the year's work provides the opportunity to report and to reflect. This year-end communication includes a portfolio (described below), a poster, and a presentation. The conclusion of the GPS project provides another example for communication beyond the SEF program. Each year, hundreds of students complete doctoral dissertations and yet do not publish their work for the larger community to see. This is a lost opportunity to share thorough and meaningful work. The work of the GPS project by the Fellows is not quite as in-depth as a doctoral research study, but does represent a significant amount of work that is helpful to the larger education community. Fellows are encouraged to not only present their work at the year-end poster session (a required element of the program), but also at local and national scientific meetings. Fellows should also be encouraged, with the support of the GPS mentor, to publish the project in trade and/or peer-reviewed journals to share the work with others.

Portfolio

The portfolio of the GPS project is a key way to preserve and document the GPS project for future Fellows. Portfolios have been created as either print documents or electronically. Each site has determined what is most needed in the portfolio. Example GPS portfolio structures of three sites are shown in Appendix 4C. These have been created using Wix (or another website building tool) as a medium.

> *This portfolio supported my growth in science in a vast way. To begin with, I have deepened my understanding in content knowledge and application. First, I began my fellowship studying place-based learning for students. This created more excitement in content because students felt vested in their learning. The fellowship also emphasized collaboration, which I implemented in my classroom on a daily basis between lab groups. Lastly, this year I placed focus on interactive notebooking and claim, evidence, and reasoning. This cultivated deeper scientific understanding and application. (Taylor Mislevich, Eldon R-1, Elementary School, MO)*

Poster

In addition to the comprehensive portfolio, each Fellow produces a poster summarizing one component of their GPS work. Three posters (elementary school, middle school and high school) illustrate the content and expectations of the poster. The posters are presented at the end of the year and serve as stimuli for the next cohort of teacher-leaders who will formulate their own GPS projects for the coming year. Many Fellows hang their GPS posters in their classrooms (Figures 4.1–4.3).

Conclusion

Over the past two chapters, we have described the two years of the SEF for Fellows. In the coming chapters, we will provide additional details on the roles of the DSC and the IHE staff. Reflections by Fellows can summarize the value of the second year of SEF.

The true impact of the Fellow as a teacher-leader is not seen during the GPS year but in the years following this intense, guided work. We have seen some Fellows who have used the GPS as a springboard to continue their work as teacher-leaders. One Fellow (a 1st grade teacher) leveraged their GPS project into a maker space for the entire school, supported by a $60,000 grant from Lowe's. Another Fellow's book project on middle school science fairs inspired a Fellow from across the country to create a book on recycling. Another Fellow began work with educators in Vietnam implementing the same lessons and activities. Fellows in multiple schools created co-teaching models that then

Figure 4.1 Elementary school example of GPS poster.

Source: Image by Karma Paoletti.

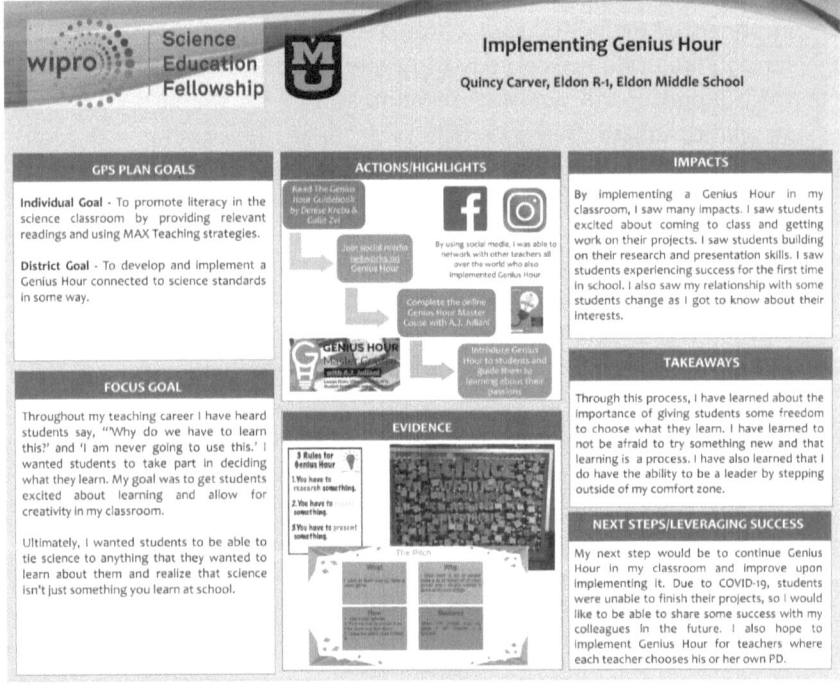

Figure 4.2 Middle school example of GPS poster.

Source: Image by Quincy Carver.

A Roadmap for Transformative Science Teacher Leadership

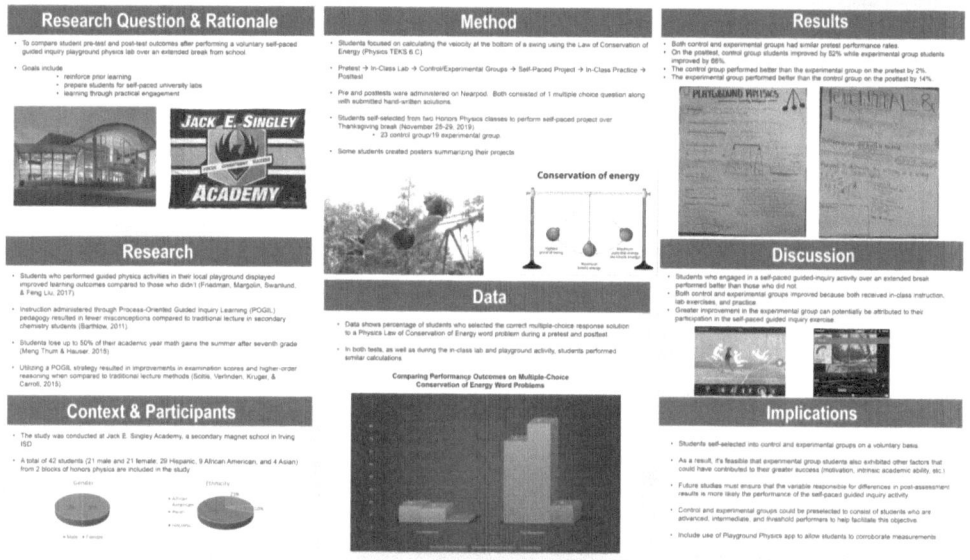

Figure 4.3 High school example of GPS poster.

Source: Image by Juan Morel.

assisted a STEAM day. One Fellow was awarded a STEM grant from the Society for Research and the Public for her work with students and leadership.

> My teaching "toolbelt" had definitely been expanded in the past two years. One of my biggest takeaways is an expanded, more in-depth understanding of the NGSS science and engineering practices. I was impacted through the research-based articles, the "practice" lessons completed by my V-CCLS and H-CCLS teams. As well as the information and real-life experiences that were shared through the various presentations that I attended (viewed).
>
> I now feel both confident and passionate about making sure to provide practice and focus on these practices for my students. (Amy Bartlett, Hallsville R-IV, Elementary School, MO)

> As I reflect over the last 2 years with [SEF] and what I did, my GPS plan will have the greatest impact on my teaching in the future. In learning about different online resources and then beginning to use them, I opened a whole new area where I will be able to impact the learning of students. I have been for most of my career a teacher

that used technology when I needed it ... This past year with Covid, the students having a school-issued device was paramount to the learning, but it meant my old ways of teaching had to adapt. I had to switch my lessons from paper to computer. This meant either finding digital versions or adapting lessons to a new format. I will continue to use the digital lessons and the online resources I utilized this school year and how I teach will be forever changed. To better serve my students, I need to help them develop the skills to be in this digital world. (Seth Willenberg, Columbia Public Schools, High School, MO)

From looking at my first reflection from this year compared to now, I have grown so much and learned so much. I have never focused on something in my teaching for an entire school year like I have for my GPS. It was very beneficial being able to find new tech resources to use in my classroom and to also tie in with our Chromebooks that my district received this year. It has been great finding so many new resources I can use in my classes and also being able to share those with other teachers in my district. I also liked the fact that I have made so many new teaching friends through this program. We have been able to keep in contact, ask each other questions, ask for advice, and bounce ideas off of each other. (Erin Snelling, Hallsville R-IV, High School, MO)

References

Doran, G. T. (1981). There's a S.M.A.R.T. way to write management's goals and objectives. *Management Review, 70*, 35–36.

Taylor, M., Goeke, J., Klein, E., Onore, C. & Geist, K. (2011). Changing leadership: Teachers lead the way for schools that learn. *Journal of Teaching and Teacher Education, 27*(5), 920–929.

Taylor, M., Klein, E. J., Trabona, K. & Munakata, M. (2022). Feminist teacher leadership: Disrupting the patriarchal binary. In N. Bond (Ed.), *The power of teacher leaders: Their roles, influence, and impact* (pp. 213–225). Routledge.

Wenner, J. A., & Campbell, T. (2017). The theoretical and empirical basis of teacher leadership: A review of the literature. *Review of Educational Research, 87*(1), 134–171. https://doi.org/10.3102/0034654316653478

York-Barr, J., & Duke, K. (2004). What do we know about teacher leadership? Findings from two decades of scholarship. *Review of Educational Research, 74*(3), 255–316. https://doi.org/10.3102/00346543074003255

Appendices

Appendix 4A: Template for writing and revising GPS goals

Name	District	
Advisor	Cohort	
Title of GPS		
Goal 1		
Goal 2		
Background/Motivation and Vision: Provide a brief overview of the background and motivation for this GPS. What is your vision for your GPS?		

For Each Goal provide the following information		
Goal #: *During the _____ school year, I will*	**Is this a district or personal goal?**	
What is your plan to achieve this goal? *To achieve this goal, I will …* **What are the** <u>actions</u> **you will take to meet the goal? List actions in the timeline below.**		
In what timeframe **will key actions be completed and benchmarks achieved? 100 hours total for all goals.**		

Date	Action	Time Estimate in hours
By this date...	*I am going to do this....*	*It will take me this many hours*
Percentage of overall hours for this goal	20/100 = 20%	20%

How will you <u>measure</u> **your progress toward meeting this goal?** *I will measure my progress by* What evidence will you provide?
What help/support (financial or other) do you need to complete your goal? This should be aligned to your actions. Include a budget of how you will spend your $1000. All purchases must receive prior approval.

Goal #		
Item(s)	**Rationale for this purchase** How will this purchase help you to achieve your goal?	**Estimated cost**

Appendix 4B: Sample of Year 2 timeline and schedule (each schedule is site specific)

Name of activity	Description of activity	Target date for completion
Draft of GPS to coordinators	Have a completed <u>1st draft</u> of your GPS and send an electronic version to **your district coordinator (DSC), the** SEF mailbox.	August 16
GPS feedback back to Fellows	Feedback will be provided by SEF leadership, and your District Science Coordinator (DSC).	August 30
Discussion with principal	Review plans for the year. Have a conversation with your principal about your goals for the GPS.	September 16
Assignment of advisor	Set-up meetings with advisor and share GPS with advisor.	September 16
Submission of revised GPS plan	Revised GPS proposal following discussion with principal, DSC and feedback from leadership.	September 30
Attend cohort meeting	Share progress with other Fellows.	October 6
Submission and acceptance of final plan	You and your advisor agree on the final plan and you submit a final version.	October 16
1st quarterly report	An account of work accomplished to date and a summary of meetings with your Advisor. Send to SEF mailbox, and cc your advisor and DSC.	November 15
Payment 1	Revised GPS, 1st quarterly report received; and Fellow has met with advisor at least once.	Stipend, in-service credits
Attend cohort meeting	Share progress with other Fellows.	Feb. 9
2nd quarterly report	An account of work accomplished to date. Send to the SEF mailbox and cc your advisor and district coordinator.	Feb. 15
Payment 2	2nd quarterly report received.	Stipend, in-service credits
Attend cohort meeting	Share progress with other Fellows.	March 2
Attend cohort poster session	Cohort Fellows will create a poster of their GPS work accomplished by this date. Audience of invited guests and former Fellows.	May 11
Payment 3	Present at Poster Session on May 11, 2019.	Stipend, in-service credits
3rd quarterly report due to advisor	Submit 3rd quarterly report to the SEF mailbox and cc your advisor, your university lead, and your DSC. Show satisfactory progress towards goals.	June 15
Payment 4	3rd quarterly report received.	Stipend, in-service credits

(Continued)

Name of activity	Description of activity	Target date for completion
Wrap up meeting with advisor	Check in	Summer
Video 1 PD session with others	PD session may be done during the school year or during the summer *but before August 1.*	August 1

Table 4B.1 Example of benchmarks and meeting templates for GPS year

September

Item	Description	Date of completion:
Choose, draft, and finalize GPS project description and goals	Complete first draft of your GPS Planning Template\and upload your template to GPS Project List – Cohort 3 by October 8, 2021	
Monthly mentor check-in	Meet with your mentor via phone/zoom during September 2021 to discuss and finalize your GPS project idea.	

Notes from mentor meeting:

October

Item	Description	Date of completion:
Discussion with principal	Meet with your principal/admin. to discuss your SEF goals and GPS project.	
Monthly mentor check-in	Have a clear timeline set for the actions you will take for your project and be prepared to discuss what you have done so far with your mentor.	
GPS portfolio	1. Open your Wix account and familiarize yourself with the platform. 2. Complete SEF GPS column (Copy/link your template to the appropriate parts of your portfolio) Description of project Goals Timeline Budget/Resources	

Notes from mentor meeting:

(Continued)

November and December/January and February/March and April

Item	Description	Date of completion:
Monthly Mentor Check-In Reflection Journal and Artifact #1	Discuss what you have done, where you are on your timeline, what you might need help with, etc. 1. Complete Journal Reflection #1: Write a one-page reflection using the prompts on the portfolio. What have you accomplished so far? What was successful? What was challenging? What surprised you? What questions do you have? What will you do next? 2. Upload Artifact #1 with a description of the artifact. (Follow the prompts on the portfolio.) Artifacts can be anything from images, documents, PowerPoints, agendas, student work, etc., that relate to your GPS project.	

Notes from mentor meeting:

Notes from mentor meeting:

May

Item	Description	Date of completion:
Monthly mentor check-in	Discuss what you have done, where you are on your timeline, what you might need help with, etc.	
Create GPS poster for end-of-year conference	Details TBD	
Create GPS presentation for end-of-year conference	Details TBD	

Notes from mentor meeting:

June

Item	Description	Date of completion:
Monthly mentor check-in		
Complete the Pillars portion of the portfolio	Reflective Practice Adult Learning Teacher Leadership Continuum	
Submit portfolio by June 30, 2022		

Notes from mentor meeting:

A Roadmap for Transformative Science Teacher Leadership

Appendix 4C: Example GPS portfolio requirements from three different sites

CA	TX	MO
Home page	Home	Home
Brief bio	About me	Brief bio
GPS project idea	Educational philosophy	GPS plan
What I hope to accomplish	SEF	District goals (SMART goals and plan-for goal)
SEF GPS page	District goal	Personal Goals (SMART Goals and plan for goal)
Description of project	Specific and strategic goal	Budget
Goals	Measurement	Reflection journal
Timeline	Actions	Instructions: In your monthly reflection journal, you should address the following topics each month. As you address these topics, you can include things that happened in your classroom and your meetings with your advisor. Be sure to refer to the Measurable, Action-Oriented, and Results/Progress segments of the "SMART" plans you defined for your GPS plan. Include images, photos, and video snippets if you wish.
Budget and resources	Results focused evidence	Describe the progress have you made toward your District Goal (include evidence).
Reflection journal page	Time frame	Describe the progress have you made toward your Personal Goal (include evidence).
Instructions	Reflections	What are the wows and yikes you've had this month?
For each journal entry, reflect on what you have done with your GPS project so far.	Data	Monthly reflections, October – June
1. Describe all of the actions you have taken for this time period.	Required help and support	Final Reflection
2. Describe what you think went well.	Budget	
3. Describe the challenges that you encountered and how you addressed these challenges.	Project resources	Quarterly reports
4. Were there any surprises that happened and/or A-HA moments?	Personal Goal (with same structure as District Goal)	Meetings
5. What questions do you have at this time? What support do you still need?	Informal Goal (with same structure as District Goal)	Dates of all meetings

(Continued)

CA	TX	MO
Reflection #1	Leadership and Innovation Goal (with same structure as District Goal)	Participants at the meetings
Reflection #2	Grant proposal (including budget and justifications)	What transpired at the meetings
Reflection #3	Professional development	Updates on progress
Reflection #4	Conference presentation	Wows and Yikes for this quarter
Final Reflection	Feedback from others	Other pages
Artifacts and Evidence page Instructions	Feedback reflection	Artifacts and evidence
	External Conference presentation(s)	Impact and leverage
For each entry, upload artifacts that provide evidence of the work you did and the growth you experienced for each of your goals (personal and district). These might be artifacts from a professional development session you gave, notes you took at a workshop/meeting, agenda for a presentation, PowerPoint slides, student work, photos of your work, etc.	Reflection journal	Adult learning
Explain how each artifact addresses the goals of your GPS project.	Successes	Teacher-Leader continuum
Explain why you chose each piece and how it demonstrates your growth toward achieving this goal. Identify successes and roadblocks, and how you dealt with each.	Opportunities for growth	Professional development
Artifacts and Evidence #1	Lessons	
Artifacts and Evidence #2	Other pages	
Artifacts and Evidence #3	Impact and learning	
Artifacts and Evidence #4	Adult learning	
Artifacts and Evidence Final	Teacher leadership	
SEF Pillars	Cover letter	
Reflective Practice	Feedback	
Adult Learning and Professional Growth		
Teacher Leadership		

Chapter 5

The District Science Coordinator

Arthur Eisenkraft and Larry R. Plank

I'm super grateful for my Wipro community.

Working with the K12 Wipro Science Education Fellows from my district as the District Science Coordinator (DSC) in San Francisco Unified School District has been an incredible experience. The teachers I work with have taught me so much about what it means to be in my district and how we can and should continue to learn more throughout our educational career.

One of my first years as DSC was during the 2020–21 school year. We were just ramping up our work when we moved from in-person classes to Zoom classes due to the Covid-19 pandemic. The teachers in this cohort were tested in so many ways and yet still created some incredible GPS projects that continue to have impacts on their teaching and their schools.

My goal during this time was to keep my teachers engaged and supported with their Wipro projects amidst the uncertainty and anxiety of online learning.

Incredibly, I found so much joy and camaraderie with these teachers – even in our online format. We were going on virtual hikes in our own neighborhoods and showing off our views, baking together online using shared recipes, and even playing networked games together (that often translated into fun Zoom classroom activities).

In our conversations, we were constantly coming up with and sharing amazing innovations and resources with each other. These moments were some of my favorite educational experiences during the craziness of the pandemic. And, I shared many of these successes with other DSCs in our monthly meetings since we were all grappling with ways to keep up the morale among our overly stressed teachers.

The learning that I've experienced at Wipro has only continued to grow in the years since Covid. Stanford – our institute for higher learning – treats us all so well. Our meetings are filled with incredible readings and activities, catered with incredible food, and make us all feel like we're part of an even larger educational community.

I've continued growing my own leadership skills through collaboration with my teachers and other DSCs, and have been able to reuse many activities and readings from Wipro in my work with other teachers in my district. Even though I'm one of the least experienced partners in the Wipro community, I've been able to co-lead sessions for DSCs from across the nation. It's been so much fun to learn from and with our colleagues throughout the Wipro network.

I know that every time I get together with our Wipro Fellows in person or online, with fellow DSCs from my area or from across the county, that I'm going to have some rich discussions about solutions to the challenges that we face as educators. Yes, we are all growing our leadership skills, whether for our science classrooms or for our respective district positions. But, more importantly, we're creating and nurturing a network of educators that prize high quality science education in our schools and will stop at nothing to leverage resources to help make that happen. The Wipro family is full of solution makers - and I'm thrilled to be a part of it. I know that this is a community that I will stay connected to for my entire career and beyond. (Eric Lewis, SFUSD, CA)

Adding the work of a SEF program to an already overwhelming workload seems like it wouldn't be worth the commitment. Having lived the experience, I know all of the wonderful ways this program enriched not only the teachers, their students, and the district – but my work and my life!! (Pam Pelletier, Boston Public Schools, MA)

In prior chapters, we have presented a compelling case for the Science Education Fellowship (SEF). We have explored the overarching goal of the program (district transformation through teacher leadership), its basis within teacher leadership literature and theory, and taken a deep dive into the program itself. We have highlighted the benefits and strengths of utilizing a distributed leadership approach (e.g., Spillane et al, 2001) and discussed the roles of formal leaders such as site-based administrators, district-level administrators, and curricular or content leadership within science at the district level, such as supervisors, coaches, and what some districts may refer to as coordinators. In these roles within districts, science curriculum or content leaders have widely varying job descriptions and responsibilities. Some of these depend on the size of the school district, the number of schools and teachers, and the grade levels that the science content expert manages. They often take a lead role in developing and selecting curriculum and instructional materials, purchasing equipment and supplies for science learning, supporting classrooms by observing and providing feedback on instruction, and assisting with local and statewide assessments (Whitworth, Maeng et al., 2017). In addition, science leaders within a district provide professional development, instructional coaching, and serve as liaisons between science teachers and site-based principals and other district-level administrators (Figure 5.1).

In our SEF program, each participating district designates a District Science Coordinator (DSC). This is typically the district's content lead. DSCs are instructional experts with formal leadership titles. This is different than the Fellows and other teacher-leaders who are serving as instructional experts without holding a formal leadership title. For our purposes, we make the distinction that the DSC has a title and official position and is quite different from the teacher-leaders who have no such positional power. This is in line with current research in the field (e.g., Bateman et al., 2024).

Within the project, the DSCs become an even greater resource to their districts as they work with and learn from university faculty, teachers, and science leaders from other neighboring districts, as well as additional Wipro SEF sites across the country.

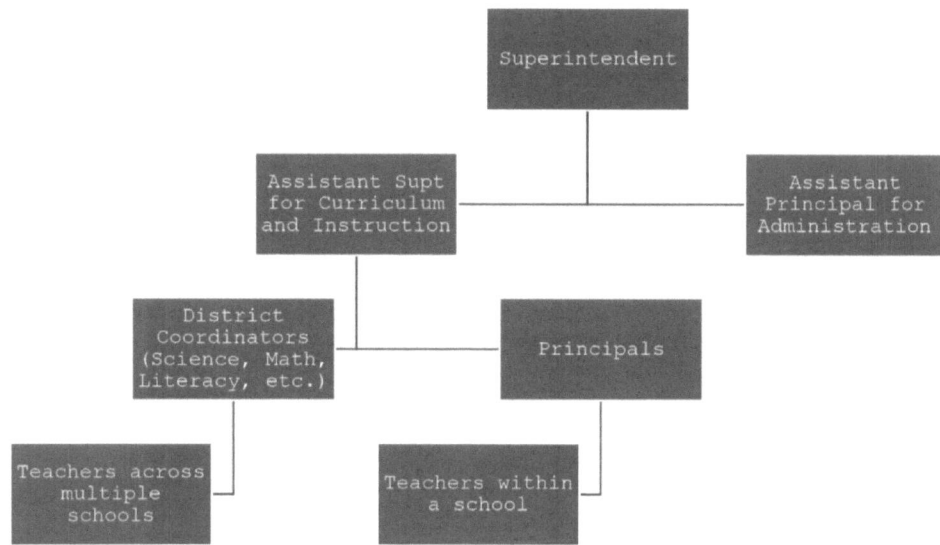

Figure 5.1 A simplified organizational chart for a school district.

During their involvement in SEF, DSCs develop their leadership capacity. Our evaluation has found that the SEF DSCs have also grown in their ability to capitalize on the teacher-leaders that the program develops in their districts in many unique and fruitful ways, including training other teachers, undertaking projects that bring enrichment to their school and/or districts, and tackling challenging reforms needed in the school and/or district. In the following sections, we will explore the role of the DSC in the SEF program as a *Collaborator, Advocate, Recruiter, Facilitator, Mentor, Manager, Learner, and Leader*. This is similar to how Heredia and colleagues (2023) explore the role of science teacher-leaders.

Table 5.1 summarizes the main responsibilities of the DSC. Additional detail is provided in the paragraphs that follow.

Table 5.1 The many roles of District Science Coordinator

Collaborator	Advocate	Recruiter	Facilitator

• Key point of contact and collaboration with university partner.
• Fosters relationships with teachers, administrators, and external partners.
• Collaborates with other DSCs in the recruitment of Fellows.
	• Develops awareness of the project within the district.
	  ○ Shares project with school board, cabinet, and leadership.
	  ○ Facilitates meetings with university and district leaders.
		• Develops strategy for recruitment of Fellows.
		• Develops recruitment materials for Fellows.
		• Assists prospective Fellows with their applications.
			• Facilitates meetings and professional development with Fellows.
			• Assists in the hosting/planning of university meetings with Fellows.
			• Encourages Fellows to attend regional and national meetings.

(Continued)

Table 5.1 (Continued)

Mentor	Manager	Learner	Leader

• Mentors Fellows during the GPS year.
• Suggests topics.
• Helps to align GPS with district priorities.
  ○ Reviews GPS proposal with the Fellow.
  ○ Develops support for the GPS from principals, coaches, and department chairs.
	• Manages the marketing program for Fellow recruitment and selection.
	• Develops and sustains communication between Fellows and university leaders.
		• Participates in a professional learning community (PLC) with other DSCs.
		• Learns alongside and with Fellows.
		• Attends and presents at leadership conferences.
		• Learns more about university and community resources.
			• Engages with the broader science education community to stay current in best practices in science teaching.
			• Models distributed leadership strategies in all that they do.
			• Supports teachers and school leaders in ways that empower them to do their best work with children and families.

The District Science Coordinator as a collaborator

The role of collaboration in an educational setting is not just beneficial but transformative, weaving through the many layers of a school district's structure to create a vibrant tapestry of growth and learning. The DSC can be at the heart of this dynamic process by actively fostering positive, productive relationships among teachers, administrators, and external partners.

Collaboration within an educational workspace is dynamic and occurs within and between many levels of an existing hierarchy of a school district. These occurrences may include sharing information, expertise, observations and reflections; overcoming territoriality; instilling a community-wide expectation of ongoing reflection and professional development; participating in co-planning and co-teaching; working to improve communication, developing a sense of belonging and membership in a learning community; and creating a common vision/a shared purpose (Bateman et al., [In Review]; Whitworth, Bell et al., 2017; Whitworth, Maeng et al., 2017). In our project, the DSC is central to developing collaborations in all these ways.

In the SEF program, we strive for collaborative inquiry where all participants (DSC, Institution of Higher Education [IHE], Fellows, principals, school administrators) work as "fellow professionals" (Hargreaves & O'Conner, 2018).

The DSCs partner with university faculty through the mutual development of program goals. Together, they ensure that the program will meet the needs of the teachers and districts. They help bridge the academic research aspects of science education with the practice of teaching science in the schools. A deep collaborative relationship between the university faculty and each DSC is key to the success of the SEF partnership. It is critical that the university faculty and the DSCs work together to develop shared programmatic goals, and that the university faculty support the DSCs as they manage the relationships within the program at each level in the district, market the program to district leadership, and develop a recruitment plan. Often, a university may not have a strong relationship with district leadership, and the DSC becomes the conduit for this to take place from the beginning of the project.

The DSCs collaborate with the teacher Fellows in the program. They encourage and support the sharing of lessons that the teachers of the program undertake in Year 1. They help shape the teacher initiative projects that form the backbone of Year 2. They build collaborative relationships during the district meetings. When Fellows interact with their DSC and view the DSC as a Collaborator rather than an evaluator or supervisor, the dynamic of their relationship shifts in a positive way. The DSCs also collaborate with their school administrators. They are active participants in shaping district priorities and initiatives. They ensure that the teachers with whom they work have a context through which to understand the spirit of new district and school programs.

Together, DSCs navigate the intricate relationships within the district, effectively promote the program to their teachers and leadership, and devise strategic recruitment plans that serve across all of the districts. The DSC plays a crucial role in bridging the gap between the university and their Fellows, often serving as the linchpin that initiates and sustains this vital connection from the outset. Through their dedicated efforts, the DSC not only enhances the program's visibility and viability but also ensures its alignment with each of the district's educational objectives, marking a significant stride towards achieving collective educational excellence across all districts.

The District Science Coordinator as an advocate

An important role of the DSC is to ensure that all levels of the district hierarchy are aware of SEF's programmatic importance, benefit, and structure. As these elements evolve, and leadership changes within a district over the life of the project, the DSC communicates and *advocates for the program* with personnel to continue their support at all levels and serves as a "cheerleader" for the project. Like many workspaces and places, school districts are not without political minefields. It is not a rarity for a well-intended program or project within the K-12 system to fail due to communication gaps as the framework is developed or project implemented. These gaps can occur with or between the superintendent, the cabinet, other district leaders, site-based administrators, and of course, teachers. Proactive advocacy for any program helps ensure its success.

Examples of serving as an advocate include, but are limited to: meeting with cabinet-level and site-level administration to ensure all levels are knowledgeable of the programmatic goals; facilitating meetings between university and district leadership; ensuring school board members are aware and sharing information publicly at a school board meeting; and developing a comprehensive communications plan for internal and external purposes. We have learned that for *District Transformation* to have the best chance of taking place, administrators at levels – but especially site-level – must have a sense of "ownership" of the program. For this reason, the advocacy of a DSC is extraordinarily important.

> *Strategies for setting up and keeping relationships between Wipro Fellows (programs) and district administration are decided by the size of the district. Smaller districts tend to have a clearer line of sight between campus operations and central office. Larger districts have more layers to navigate. Fellows thrive with administrative support at all levels, from the campus to the central office administrator who signs the legal agreements between Wipro and school districts.*
>
> *The DSC works to ensure all district stakeholders are aware of the positive impact the SEF relationship can offer public schools. Creating opportunities for administrators to see the Fellows in action is beneficial. (Leading Professional Development sessions, writing curriculum, working with students, presenting at professional conferences.) These wonderful things happen because of the contributions Wipro makes to the development of Science Education Leaders.*
>
> *Chris Dazer, Director of Science Discovery Education, Irving ISD, TX*

The District Science Coordinator as a recruiter

The DSC plays an integral role in the process of recruiting teachers who show promise, want to lead from the classroom, and desire to play a role in transforming their school site in the spirit of influencing district transformation. For some DSCs, the recruitment is seen as an opportunity to offer to teachers something that could make them feel respected and professional.

While we know the program is extraordinarily valuable and transforms the careers of those involved, with a goal of leading to district transformation and student achievement, the selection of Fellows can be a difficult process. Teachers are continually overwhelmed in their duties, and, very often, they elect not to participate in additional growth opportunities due to stress levels, and/or an attitude of "one more thing" being added to their plate. The recruitment process can assist teachers in surmounting these hurdles.

In our program, districts have four Fellows each year for three years, for a total of 12 Fellows across the three cohorts of teachers. This results in an annual cohort of 20 Fellows across five school districts associated with a particular university each year. These numbers may vary in the case of very large or very small districts (i.e., rural districts). The DSCs of the districts worked together to balance the recruitment across grade levels throughout the life of the project not only within their own districts, but also across all the districts in the entire project. The task of selected recruitment across districts and grade levels becomes more challenging over time, yet this balancing of future Fellows is crucial for district transformation as shown in Table 5.2. Therefore, ongoing attention to recruitment is a significant task for the DSC.

During recruitment, the DSC may target specific teachers to apply but should also open up the application process to all teachers who meet the requirements of becoming a Fellow (e.g., three years in the district, committed to remaining in the classroom for at least two years post award). The opportunity to become a SEF Fellow should not be limited to those teachers who have already demonstrated teacher leadership but should also include those who wish to grow as leaders. The DSC should be available to assist teachers with the application process. This may include proofreading an essay

Table 5.2 Sample of recruitment of teachers across districts over time

	District A	District B	District C	District D	District E	Annual Project Totals
Year 1	2 ES, 2 HS	2 ES, 1 MS, 1 HS	2 MS, 2 HS	3 ES, 1 MS	1 ES, 1 MS, 2 HS	8 ES 5 MS 7 HS
Year 2	2 ES, 1 MS, 1 HS	2 MS, 2 HS	3 ES, 1 MS	1 ES, 1 MS, 2 HS	2 ES, 1MS, 1 HS	8 ES 6 MS 6 HS
Year 3	3 MS, 1 HS	1 ES, 1 MS, 2 HS	1 ES, 1 MS, 2 HS	2 ES, 2 HS	1 ES, 2 MS, 1 HS	5 ES 7 MS 8 HS
Project district totals	4 ES, 4 MS, 4 HS	3 ES, 4 MS, 5 HS	4 ES, 4 MS, 4 HS	6 ES, 2 MS, 4 HS	4 ES, 4 MS, 4 HS	21 ES 18 MS 21 HS

Abbreviations: ES = Elementary School, MS = Middle School, HS = High School.

or serving as a thought partner as a potential Fellow is drafting a statement of interest. Maintaining clear records of applicants over the three-year project is important, and this may also allow for a "deepening of the bench" over time to further recruit for subsequent years. Our most successful DSCs never underestimated how personal, direct connections to all applicants, whether they are selected or not, can make a world of difference.

When the due date for applications arrives, the university will set up a meeting where the university leadership and the five DSCs from participating districts help choose the Fellows for Cohort 1. Using the agreed upon criteria for selection, applications are reviewed, and the potential Fellows are selected. The DSC should encourage applications from their teachers but should not choose the Fellows from their own district. This is done to avoid any conflicts of interest or perceived conflicts of interest. For example, Fellows not chosen can accuse the DSC of favoritism or conjure up rationale for why they were not chosen having to do with personal bias. By recusing oneself from the selection of the Fellows from their own district, potential appearances of conflict can be avoided.

Once the Fellows were chosen, DSCs celebrated the process and programmatic requirements with communication to all levels of district staff, the school board, and the public.

The Process of Recruiting Teachers for the Wipro SEF

We began to recruit teachers by holding meetings after school to introduce interested teachers to the Wipro Science Education Fellowship. At our meetings, we discussed the purpose and goals of the fellowship, and what the requirements would be to become a part of the Cohort.

A member of the leadership team from Montclair State University attended our meetings, providing crucial support to move the recruitment process forward.

As a result, we were successful in recruiting teachers to fill the places we had been allotted. Once the cohort was established, our leadership team held regular meetings to provide professional development and a forum for these teachers to meet and discuss their progress. The cohort also had the support of the District Coordinator, who acted as a liaison between the teachers and the leadership team.

The important thing to remember when recruiting new Fellows is to seek out teachers who love science, are willing to help their colleagues, who want to be teacher-leaders and are in this for the benefit of the students. These individuals will be the builders of a strong fellowship. They are the ones who will help continue its success and move forward with implementing new ideas in the future. As teacher-leaders, they will be the ones who help their colleagues implement NGSS-aligned instruction in their districts and establish groups of teachers who learn from each other through collaboration and sharing.

Beyond the initial cohorts, willingness to continue their work through additional phases of the fellowship is important. These Fellows will recruit other like-minded teachers by inspiring them to participate in their ideas and grow with them. The beauty of the program is the ability to be able to recruit others who have a love of science in their hearts and the desire to help and inspire others. These are the individuals who embraced becoming a part of the Wipro family and believed that "once a Fellow, always a Fellow!"

Mary Goffredo, District K-12 Math and Science Supervisor, Kearny School District, NJ

The District Science Coordinator as a facilitator

The role of facilitator by DSCs varies based on the type of meeting they were engaged in and across the duration of the program.

During the first year and throughout the program, the DSCs attend three types of meetings: program-level meetings (monthly) with the cohort of 20 Fellows organized by the University; district-level meetings (at least quarterly) with their selected Fellows organized by the DSC; and project-level leadership meetings comprising DSCs and university faculty. The district-level meetings help form a bond among the DSCs and Fellows. This bond will be important in the second year where Fellows are working on their GPS.

At the start of each site's program, the task of facilitating the monthly meetings of the cohort Fellows was shouldered mainly by the university faculty. In some cases, after a few meetings, the DSCs were invited to co-facilitate with university faculty or lead portions of the meeting to share.

The DSCs play an important role in modeling practices, facilitating learning-based meetings, and hosting/facilitating leadership meetings with the university. During these experiences, our DSCs developed additional leadership attributes and became better able to facilitate professional development within their districts, conference sessions, and other site-level and district meetings.

Regular monthly meetings with Fellows in the Tampa Bay area have played a pivotal role in nurturing science teacher leadership across our three districts. Historically, opportunities for science educators to collaborate and exchange best practices beyond their districts have been limited. The Wipro fellowship has bridged this gap, empowering our teachers to glean insights from one another in diverse ways. Structured monthly meetings enable teachers to showcase successes and address challenges collectively. Furthermore, our monthly journal club facilitates the sharing of research findings on various topics in science education. This interactive platform not only fosters engagement with research but also allows for meaningful discussions on its practical application and exchange of implementation ideas.

Fawnia Schulz, Pinellas County Schools, FL

The District Science Coordinator as a mentor

During Year 2 of the SEF program, Fellows began to work on professional development plans referred to in our project as a *Growth Plan System* (GPS). During this process and throughout the second year, the DSC served in a much more active and important capacity in personally interacting with Fellows and serving as mentors and coaches in their work.

Our DSCs typically had unique insight into district-level initiatives that our Fellows generally did not have in working day-to-day at a school site. Therefore, our DSCs suggested ideas and collaborated with their Fellows all the while making suggestions to potentially align the project initiatives of their respective school district and school site. Often, DSCs assisted in planning meetings of Fellows with principals, academic coaches, grade-level leaders and/or department chairs to ensure the Fellows had the support of their site-based peers and colleagues.

Once the general form of the GPS takes shape, the DSC can lend support to the Fellow as the GPS is discussed with the school principal. It is best if the principal is seen as a partner in the formulation of the GPS and not merely as a final signatory. Principals have a school perspective that can strengthen a GPS if they are part of the conversation. Support from the principal can also help make involving other teachers and assets of the school (materials, conference rooms, etc.) in the project easier. The discussions regarding the GPS can form a new avenue of dialog between the principal and the DSC, allowing a larger conversation of what is valued in high quality science education to take place. The final GPS can then be agreed upon by the Fellow, the DSC, the principal and the IHE.

As described in Chapter 4, Fellows often develop ambitious plans for their GPS and realize that executing them requires more time than they anticipated. The DSC and university team can assist the Fellow in editing their plans and acknowledging that these edits are a normal part of the process and not a signal of failure. It is through this process that the DSC helps the Fellow lean on the Adult Learning pillar of the SEF.

During Year 2, the Fellows can have time during the district meetings, led by the DSC, to discuss their GPS and progress. This group has built a community of trust and partnership during Year 1. This community both encourages and allows Fellows to feel empowered to push themselves and take risks. Hearing from the other Fellows in their district and giving and receiving warm and cool feedback is part of the growth

Table 5.3 District DSC meeting attendance through the years of the SEF

Year of program	Year 1	Year 2	Year 3	Year 4
Fellows attending district meetings	4 (cohort 1 – CCLS)	4 (cohort 1 – GPS) 4 (cohort 2 – CCLS)	4 (cohort 1) 4 (cohort 2 – GPS) 4 (cohort 3 – CLLS)	4 (cohort 1) 4 (cohort 2) 4 (cohort 3 – GPS)

process. DSCs are also capturing learnings that they can use to assist the next cohort of Fellows as they move through the GPS phase. There may also be an opportunity for the principal to invite Fellows to describe what they are doing at a school faculty meeting to generate interest and possible support.

Year 2 of Cohort 1 coincides with Year 1 of Cohort 2. While Cohort 1 is working on the GPS, Cohort 2 is moving forward with their Vertical Collaborative Coaching and Learning Science (V-CCLS) and Horizontal Collaborative Coaching and Learning Science (H-CCLS) experiences. The district meetings now include the Fellows from Cohort 1 and Cohort 2. In Year 3 of the program, another cohort joins the program, and the formal involvement of Cohort 1 is complete. Table 5.3 shows how the district meetings expand during the four years of the SEF program, building from the DSC and four Fellows to the DSC and all 12 Fellows by Year 3.

One can see how the 12 Fellows and the DSC grow as a PLC over the four years. The expectation is that they share experiences while learning and critiquing pathways for the school district (and each other) to move forward with exemplary science education. The Fellows all have experience with vertical articulation of curriculum (V-CCLS work during Year 1), the science and engineering practices of the Frameworks and Next Generation Science Standards (NGSS) (H-CCLS work during Year 1), and professional development and leadership roles (GPS during Year 2). The sheer amount of time working together is a key factor in building these relationships. When done well, the four years of participation in the SEF can be very powerful in building a cohesive, high-performing team to guide student achievement. This is a true district transformation.

> Before this SEF program, my DSC helped me better adapt to the various science teaching requirements in California public high schools because I was a new arrival to California from overseas. During SEF, I realized that the District Science Coordinator could help science teachers open their horizons to a higher level of instruction. Seeing my growth during the SEF program, she encouraged me and I could see that she was pleased. Our relationship became closer, as if we were in a team working together to accomplish something big.
>
> Yichang Liu, Gunderson High School, San Jose Unified School District, CA.

> As a third-grade teacher, I was approached by our District Science Coordinator who encouraged me to apply for this new fellowship program, the SEF. I had had an underwhelming experience with different STEM programs, so I was reluctant to apply, but

he assured me it would be worth my while. It was, and still is. As an outcome of SEF, together, we traveled to Nashville in 2016, where I presented at the NSTA Conference, and my DSC was there to support me, rescuing me when my laptop wouldn't work with the tech equipment, and helping distribute materials to the attendees. It was an unforgettable experience, and I am grateful to him.

Dave Kleiner Montclair Cohort 1, NJ

My relationship with my district coordinator feels collaborative. I feel able to suggest PD ideas and be honest with feedback about the pluses and deltas from meetings, and brainstorm/refine ideas to uplift voices from our team. Our current coordinator was not in the district when I was a participant in Wipro SEF; since then, I have implemented a similar PD model for small teams in our district. In the first meeting with our new coordinator, I shared information about the project and she immediately asked what she could do to support both recruitment and implementation. This first interaction is what set up our collaborative relationship, so SEF is instrumental in that.

Tal SebellShavit, Cambridge, MA

The District Science Coordinator as a manager

Project management focuses upon planning and organizing a project and its resources. As shown in the previous sections of this chapter, the DSC plays the role of project manager by marketing of the program, building support for and relationships among district and site-based leaders, recruiting candidates, and providing essential support to the Fellows over their two years within the project.

As a project manager in partnership with their district, the DSC is responsible for building a common vision, anticipating necessary supports, and providing professional growth opportunities for all teachers of science within the district. Through this process, they guide the formation of a professional learning community (PLC) across the district. In education, changing beliefs and instructional practices takes time and a great deal of effort. Beyond lessening the burden of the workload for district change initiatives, the DSC brings together a strong team of leaders with diversity of perspective and experience to inform decisions which should lead to better outcomes for students and thus leading to *district transformation*.

As a project manager in partnership with the university, the DSCs have forged relationships with university faculty that have led to collaboration within other projects. Many of our university leaders have shared the importance of the DSC role in the day-to-day operation of the SEF program, specifically citing the well-developed skill set of DSCs and resulting success of the program in alignment with the vision of the national leadership responsible for the program.

The goal of the Wipro program has always been to engage district science coordinators as full partners and co-collaborators in the program alongside faculty and teachers. We recognize their central role in the process of district change and, thus, finding districts with engaged and active district coordinators is key in building the program. Enacting this goal has been messy, as is so frequently the work of teacher leadership. We have DSCs who have continued to collaborate with us for a decade now, and others who provided minimal assistance (or rather were not obstructionist). Most recently we have begun to "grow our own" DSCs (former Fellows and in one instance, a doctoral student who conducted her dissertation research on Wipro before becoming a DSC), which has proven to be exciting and powerful as they have both deep knowledge of the program from the perspective of a teacher and now have the necessary leverage at the district level to support change.

Emily J. Klein, Montclair State U, NJ

The District Science Coordinator as a learner

Finally, our DSCs reported a strong sense of learning and belonging over the life of the project. This resulted in a broad community of trust and partnerships among DSCs and their Fellows, DSCs and their universities, as well as DSCs across projects in the seven sites across the United States. In science education, few opportunities exist for leadership support and development within the content area. Only one national organization, the National Science Education Leadership Association, is solely dedicated to leadership development in science education. Therefore, the larger community of practice formed by DSCs to discuss curriculum changes and instructional challenges, national and state-level legislation, district issues such as teacher retention and recruitment, and what it means to truly transform a district, played an important role in the learning and development of the DSC as leaders.

The SEF program brings five local districts together and the five DSCs have a chance to share their challenges and successes while their Fellows are moving forward in the program. The DSCs form their own PLC where issues connected to SEF and issues outside of SEF can all be discussed. Insights into how one district orders supplies or how another district does teacher evaluations or how a third district sets district priorities is relevant and important for the DSCs. The SEF program becomes a vehicle for DSC professional development by providing a common experience, a set of meetings to attend, and an opportunity to learn from one another. The loneliness and challenge of being the sole DSC in the district is somewhat ameliorated by sharing experiences with other DSCs in local districts.

The DSCs in a local context form a PLC. Another benefit of the SEF is the large national nature of the program. This allows for a significant network of talented and motivated individuals to develop, and naturally forms a second PLC in which the DSC participates. The SEF network (seven universities and 35 school districts) provides a national perspective for sharing DSC experiences and successful practices, which, in science, is a rarity, especially for smaller districts. In this context, leadership conferences are held for all 35 DSCs. This leadership conference is an excellent professional development opportunity where DSCs participate as Learners and Collaborators while also leading or co-leading sessions with university faculty or other DSCs. Sessions at recent conferences have included: Latest trends in science education, teaching controversial/politically charged topics, working with administration, successful leadership, and community building. We have had great success with these both virtually and in-person.

Here are a couple of "visual notes" from these conferences (Figures 5.2 and 5.3).

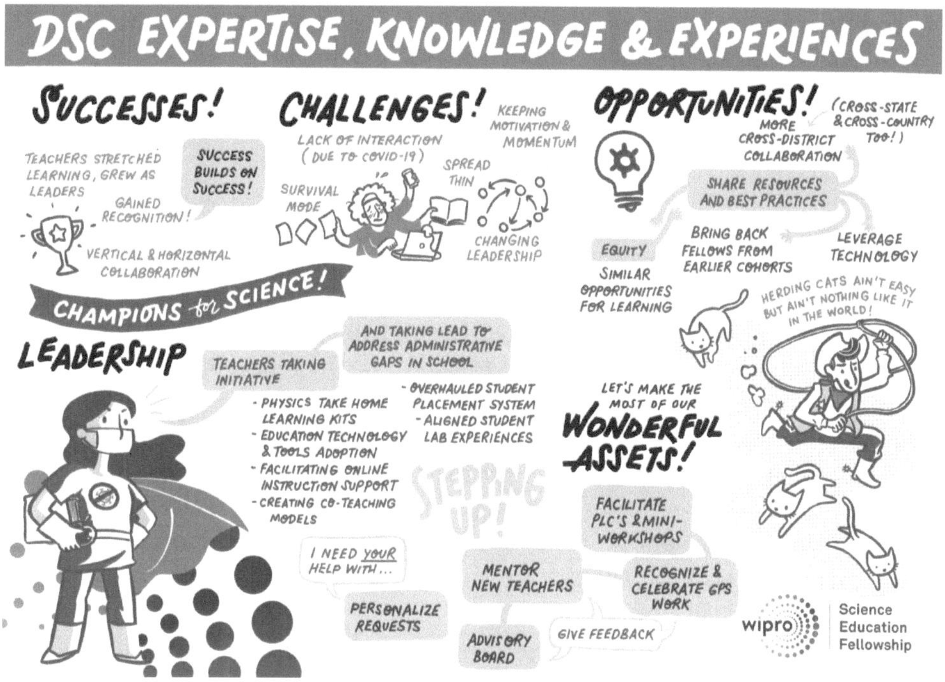

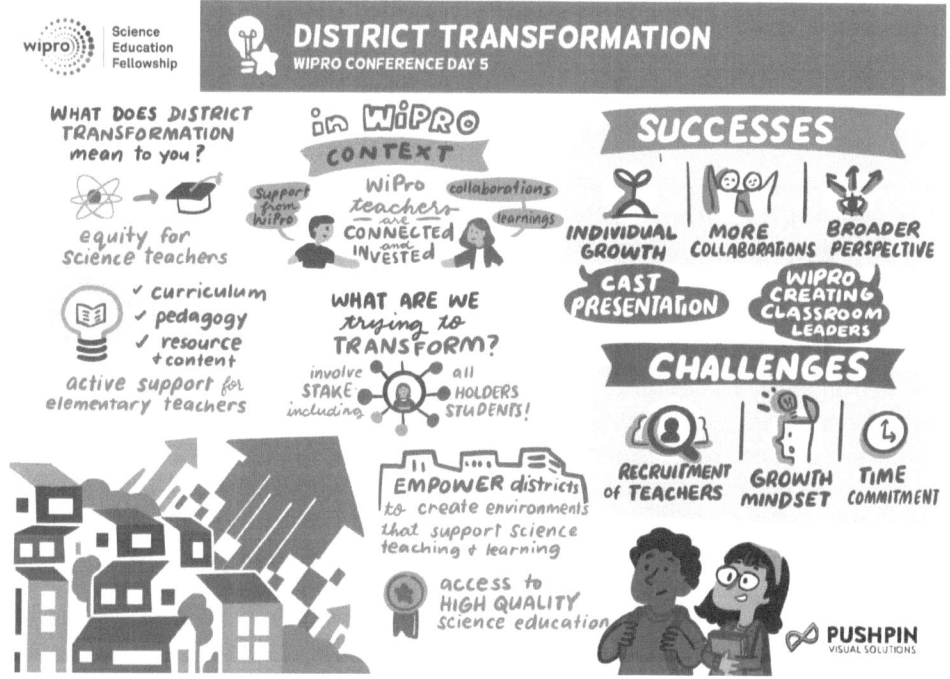

Figure 5.2 Visual notes from the February 2021 conference.

Source: Visual summaries by Pushpin Visuals.

Figure 5.3 Visual notes from the March 2023 conference.

Source: Visual summaries by Pushpin Visuals.

District Science Coordinator

In addition, through their emergent relationships with university faculty, some DSCs gained access to additional projects, events, and learning experiences at the university level and within the broader landscape of science education, both formally and informally. Since many meetings within projects were held at science-based places, such as wildlife preserves, museums, aquariums and the like, DSCs reported a stronger sense of connection to science within their communities and learned additional ways to impact science learning within their districts.

Reflecting on my three years as a Wipro Science Coordinator fills me with immense gratitude for the support and guidance of individuals like Dr. Ratana Narayan, Dr. Arthur Eisenkraft, and the district science coordinators and Fellows from partnering schools and universities, locally and nationally. Dr. Narayan and Dr. Eisenkraft offered me the invaluable opportunity to serve as a district science coordinator, enabling me to simultaneously grow in my role while providing vital support to science teachers within my district. Witnessing the impact these teachers had in their classrooms and across the district, serving as exemplary models for non-Wipro Fellows not only enhanced teaching and learning in my district but also inspired continuous growth and development in the pursuit of education.

Furthermore, the relationships forged with other UNT Dallas science coordinators have been instrumental in my personal and professional journey. Having colleagues who share my passion for science education has been immensely rewarding, as I've had the opportunity to learn from and alongside them. Collaborating on sessions with coordinators from other Wipro sites has furthered my professional growth and leadership skills.

Danielle T. Moore, TX

One facet of the learning I experienced focused on leadership. There was also a whole host of other "learning." For example, I also learned a great deal from the Fellows as teachers, as humans, as well as from all of their CCLS cycles, I learned from their GPSs, I learned from their classroom experiences. I did learn from my DSC colleagues, but mostly they were critical friends, thought partners, empathy providers, and givers-of-hugs (literally and figuratively). They eased my stresses and soothed my soul. They helped me see my strengths, identify my areas for growth, and reminded me that I wasn't in it alone. Yes, I learned things that might be helpful in my context – but, mostly – we developed a camaraderie because of our similar roles. I see that going beyond "learner" and frankly, it was what made all this "extra work" of SEF worth doing – my personal value-added (in addition to all the great things it did for my teachers, students, and district.),

Pam Pelletier, MA

The District Science Coordinator as a leader

In fulfilling all the roles listed previousuly – Collaborator, Advocate, Recruiter, Facilitator, Mentor, Manager, and Learner – the DSC exhibits leadership. The DSC motivates and enables others to contribute. The DSC learns alongside the school administration, the IHE, and the Fellows. The DSC has the opportunity to encourage a distributed leadership model in the SEF program and their district.

District Science Coordinators in the research

In educational research, DSCs are a vastly under-researched group in comparison to science teachers (Whitworth & Chiu, 2015). Additionally, few professional development programs exist for DSCs outside of the Wipro program (e.g., Bateman et al. [In Review]; Whitworth, Bell et al., 2017). Research, to date, describes the job responsibilities of a DSC and interactions with teachers and stakeholders, but these responsibilities fluctuate based on context and district demographics (Whitworth, Maeng et al., 2017). DSCs are interested in participating in job-specific professional learning opportunities, as they are often a siloed position within their district. Recent work supports using a Community of Practice framework for online professional learning to engage DSCs and increase confidence and competency of NGSS science practices and help participants feel less isolated (Bateman et al. [In Review]; Bateman et al., 2024). Given the potential impact that DSCs can have on teachers and the district as a whole, it is important to include them in projects and programs that want to see district transformation.

Conclusion

The role of the DSC is essential to the successful implementation of a SEF and much focus should be expended upon selecting great candidates within districts to be served by such a program. In this chapter, we defined the role of the DSC as a ***Collaborator***, ***Advocate***, ***Recruiter***, ***Facilitator***, ***Mentor***, ***Manager***, ***Learner***, *and* ***Leader***. As you consider and embark upon your own projects within your community, we are certain that additional attributes will emerge related to this role. You, like us, will one day celebrate your DSCs as "superheroes of science."

References

Bateman, J. M., Schwendemann, M. L., Whitworth, B. A., & Luft, J. A. (In Review). *Comparing district science coordinators' communities of practice.*

Bateman, J. M., Whitworth, B. A., & Wenner, J. A. (2024) [Forthcoming]. *In or out of the classroom? Differentiating science teacher leaders and district science coordinators: A literature review.*

Hargreaves, A., & O'Connor, M. T. (2018). Solidarity with solidity: The case for collaborative professionalism. *Phi Delta Kappan, 100*(1), 20–24. https://doi.org/10.1177/0031721718797116

Heredia, S. C., Phillips, M., Stallings, S., Worsley, T. E., Yu, J. H., & Allen, C. D. (2023). Identifying the roles of science teacher leaders in practice. *Journal of Science Teacher Education*, *35*(2), 109–126.

Spillane, J. P., Halverson, R. & Diamond, J. B. (2001). Investigating school leadership practice: A distributed perspective. *Educational Researcher*, *30*(3), 23–28.

Whitworth, B. A., Bell, R. L., Maeng, J. L., & Gonczi, A. L. (2017). Supporting the supporters: Professional development for science coordinators. *Journal of Science Teacher Education*, *28*(8), 699–723. https://doi.org/10.1080/1046560X.2017.1404814

Whitworth, B. A., & Chiu, J. L. (2015). Professional development and teacher change: The missing leadership link. *Journal of Science Teacher Education*, *26*, 121–137.

Whitworth, B. A., Maeng, J. L., Wheeler, L. B., & Chiu, J. L. (2017). Investigating the role of a district science coordinator. *Journal of Research in Science Teaching*, *54*(7), 914–936. https://doi.org/10.1002/tea.21391

Chapter 6

The role of the Institution of Higher Education (IHE)

Arthur Eisenkraft

The implementation of the Science Education Fellowship (SEF) program at universities produces benefits for the individual professors and staff at the Institution of Higher Education (IHE) as well as for the university as a whole. Creating a meaningful partnership with local school districts situates the university as an active member of the community. The work in SEF morphs the relationship of school district teachers and coordinators from participants enrolling in university courses to colleagues working together with a common purpose. They mutually benefit from a deeper relationship than the one that formed when they shared university classrooms as undergraduate or graduate students. Reading the research when you have never been in a classroom is very different than reading and using the research as an experienced teacher. The SEF structure pushes Fellows to use and apply this research. The professional learning that each IHE team plans requires teachers to think deeply about their own instruction and their roles as leaders. The teacher-to-teacher collaboration (through Collaborative Coaching and Learning in Science [CCLS], etc.) of course is important, but the critical feedback, mentoring, and advising that the IHE team gives during the construction of the Course of Study, Growth Plan System (GPS), presentations, etc., really pushes teachers to grow.

Throughout the book, we have described the program as being led by a university (IHE), which is how we have run the program. At the same time, we envision that this program could be led from a school district or by a collaborative of neighboring school districts. This would be especially true if the individual school district is experienced and savvy with science education and current research. If this is what you choose to do – wonderful! We would love to hear from you. As you read this chapter, substitute your role for the role of the Director at the lead university. If you are a university faculty,

DOI: 10.4324/9781003490586-7

or a lead at a professional learning center, then the process is written for you in this chapter.

When we were first awarded the Wipro SEF funding, we were thrilled, but really, we didn't know then how important this grant would be for us. Although Meghan and I were newer faculty, we had another couple of grants we were running, but because the SEF was this partnership across universities, we were able to see how these established colleagues leveraged their funding. The leads at the other sites served as informal role models for us in some ways, showing what was possible with funding like this. Working with the centers at UMB and MSU, it gave us the idea to start our own center – which is now in its eighth year! Also, through our research on the SEF, we found the CCLS professional development was a powerful approach, so we adapted the model in several ways in other projects with great success. The SEF generally also served as another entry point to partner with districts, opening doors for more projects and partnerships for our school and our center. The Wipro SEF was so much more for us professionally than just a funded project. It helped support the expansion of our work and research in ways we had never expected, which is wonderful.

Amanda Gunning – Mercy University, NY

The Wipro Science Education Fellowship Program is currently in its 7th year at UNT Dallas (2017 – current). The impact of the Wipro SEF program on all its stakeholders cannot be overstated! The Wipro SEF Program has propelled UNT Dallas to the forefront as a leader in Science Education in our region and cemented relationships with our partner ISDs. In a resource-depleted South Dallas, the consistent, high-quality programming offered by Wipro SEF at UNT Dallas has enabled our K-12 Wipro Fellows to grow as science teacher-leaders and agents of school/district transformation.

As the PI of the Wipro SEF Grant at UNT Dallas, in addition to the university, ISD partners and Wipro Fellows, the program has had a huge impact on me both professionally and personally. I have grown tremendously as a science educator, a leader, and as a human.

Ratna Narayan – UNT Dallas, TX

Throughout the preceding chapters, we have taken you through the layers of the program, starting from an introduction and general overview, through some theoretical background, to the heart of the program – the work of the Fellows. We have also highlighted the importance and role of the District Science Coordinator (DSC). In our model, the university played the central organizing role in implementing the program. In this chapter, we will provide insights on what we've discovered to be instrumental in the effective operation and sustainability of the program through the university or IHE.

Introduction

The IHE is the hub for the activities associated with the SEF. They are responsible for implementing all aspects of the program including maintaining local school district partnerships, working with DSCs, recruiting and selecting teacher-Fellows, allocating resources, conducting meetings, facilitating professional learning, communicating with all interested parties, planning conferences and special events, providing reports, and supporting outreach efforts. The IHE's main objective is to create a vision for excellence in science teaching and learning by deep investment in school districts and their teacher-leaders. The IHE engages the district partners in collaboratively building the norms and work overtime to ensure that the work of the districts (i.e. Fellows, science coordinators, principals, other administration), and the IHE is a true partnership with shared responsibilities, respect, and the desire to learn from one another.

The SEF is intended to sustain district transformation through teacher leadership. The program does this by investing deeply in several local participating school districts and providing professional learning experiences grounded in research to improve science classroom practice. The SEF works to build up teacher-leaders and leverage their expertise within their districts to support district science goals. The SEF provides new learning communities for teachers both within and outside of their districts. Professors and staff at IHEs form relationships with classroom teachers that allow for resource sharing and contextualized support to best meet the needs of teachers and districts.

Building positive relationships and forming a community is as dependent on displays of good will and camaraderie as it is on good administration of all details. The SEF is a large program with many moving pieces and administrative details. As with any large program, having a strong team in place to manage these details facilitates the relationship and community building needed for the program to truly flourish.

In the following sections, we will describe the necessary infrastructure as well as the roles and responsibilities that are essential to implementing the SEF by an IHE or other science education or professional learning entity.

SEF program overview (a brief retelling of information provided in prior chapters)

As the implementation of the SEF program commences, the IHE provides coordination for Year 1 and Year 2 of the program for each of the three cohorts of Fellows.

Year 1 of the program is mainly focused on teachers thinking about teaching. They participate in both their Vertical Collaborative Coaching and Learning Science (V-CCLS) groups (fall semester) and Horizontal Collaborative Coaching and Learning Science (H-CCLS) groups (spring semester). The work of the teacher-Fellows and the DSCs in Year 1 has been described in prior chapters along with some mention of the role of the IHE. The IHE creates the V-CCLS and H-CCLS teams. The IHE also supports those teams with exposure to research trends as well as securing research articles.

In addition to the team meetings that are coordinated by the Fellows, one major responsibility is for the university to run the monthly Fellow meetings. Fellows from five different school districts gather once a month at the host university to engage in professional development in the areas of instruction, reflective practice, adult learning, and leadership. The IHE works with the DSCs to plan the agendas for these meetings. The monthly meetings are held at the university or at the local school districts on a rotating basis. This depends on travel times and other logistics. Refreshments are always a welcome treat as the teachers will be arriving at the monthly meetings after a full day of teaching.

Two major conference events also take place in Year 1. The January conference provides the opportunity for the V-CCLS teams to present their work, course of study, and experiences. The June conference provides a similar stage for the H-CCLS teams. The June conference invitees include district administrators and Fellows from other sites. The June conference also has GPS posters displayed from the prior cohort of Fellows.

Planning the January and June conferences includes choosing a location, preparing agendas, enlisting a keynote speaker (June), invitations, guest lists, and food. In addition, there will be displays of the GPS posters from the prior cohort.

Year 2 of the program is built around implementing the Individualized GPS. Each Fellow develops and carries out an individualized growth plan that has a clear vision and identifiable benchmarks. The 100-hour plan focuses on ways to improve the teacher's own instruction and leadership and is developed in collaboration with a university advisor, the DSC, and the Fellow's principal. The yearlong project includes the Fellow leading professional development for other teachers in their school district and culminates with a report and presentation of a poster at the end of year conference.

During the GPS year, the IHE helps the Fellow develop the GPS plan. The IHE also matches a university advisor/mentor to support the Fellow throughout the year. The IHE will also set up approximately four meetings where the 20 Fellows can discuss their projects and progress with one another. At the close of the year, the IHE assists the Fellow in creating a poster to represent the year's work and then prints the poster. The posters are displayed at the year-end conference, where there is a chance for interactions with others.

Over a rollout of three successive cohorts of Fellows, each participating school district will have as many as 12 Fellows who have participated in the extensive two-year Wipro SEF program. These Fellows serve as a leadership group for district science and engineering initiatives. This critical mass of teacher-leaders sets the stage for district transformation.

Innovation phase

After Fellows complete the two-year "foundation" program, DSCs work with their university partners in exploring ways in which to build on the Fellows experiences, projects, and leadership skills to support district transformation. Through various and varied initiatives, Fellows engage with other teachers in their districts. Simultaneously, administrators are made more aware of the resources that the SEF program has seeded in their schools and districts. This phase of funding is also intended to encourage district incentives to support future work that will continue after external funding concludes. More will be described regarding the Innovation Phase in Chapter 8.

Table 6.1 outlines the major tasks during each year of the program after the school districts have been selected and come on board.

To get to the Implementation Phase with the Fellows, the IHE must choose school district partners and then, in turn, work with the DSCs in recruiting and choosing Fellows.

Table 6.1 Rollout of the SEF program over successive years

	Cohort 1	Cohort 2	Cohort 3	Phase II
Year 0	Recruitment			
Year 1	Collaborative Coaching and Learning in Science (CCLS)	Recruitment		
Year 2	Growth Plan System (GPS)	CCLS	Recruitment	
Year 3		GPS	CCLS	
Year 4			GPS	
Innovation Phase				Activities proposed by individual sites.

Key to yearly activities

Key roles and responsibilities: The people needed for SEF implementation

The SEF program has a complex design that requires coordination and sharing of responsibilities. In the following section, we will highlight the roles and responsibilities of key personnel needed to implement the SEF at each university. Depending on the size of the university, the other commitments of faculty, and the skill sets of staff involved, there may be variations in the titles and number of folks responsible for the tasks necessary for successful SEF implementation. Tasks and roles can be combined or adjusted if there are fewer people available for these roles. In one of our sites, the entire program has been the responsibility of a single professor with some administrative help. This is mentioned to encourage interested people to step up and not be intimidated or overwhelmed by the number and variety of tasks and sites with more limited resources.

The Principal Investigator

The Principal Investigator (PI) of the SEF program at the lead university should be a professor at that university with knowledge and positive relationships with both the College of Education and the College of Science and Mathematics. The PI sets the vision for the program and helps to secure the funding, negotiates the budget, and ensures that reporting requirements and spending limitations are met. The PI communicates with the sponsor and the necessary university hierarchy as well as oversees the implementation of the program. The PI of the SEF program gives final approval of all budget decisions and staffing and is involved in the selection of district partners.

Once a university is awarded the SEF grant, several components of the program need to be put in place for the SEF program to be implemented. The PI, working with the appropriate university departments and SEF staff, must get the budget in the university system, create a presence through university print and web media, establish the SEF staff team, and recruit five partner districts and teacher-Fellows.

The Program Manager or associate director

The SEF Program Manager (PM) works alongside the PI to ensure all elements of the program are in place. From writing requests for proposals (RFPs) to the selection of the program evaluator and the partner universities to communicating the intent of the program to district administration, the PM must have various abilities and be able to coordinate tasks synchronously. Since the PM must be able to communicate with all involved parties from sponsored programs to teacher-Fellows and district personnel, it is preferred that this person has a background in science education, teaching, and grant management.

The PM collaborates with the PI to plan monthly meetings and special events. The PM leads the recruitment effort and works with the DSCs to ensure the goals of the targeted recruitment are met. The PM works closely with Fellows from recruitment to the end of the program, providing guidance and assistance when needed. The PM will create and maintain records of all the cohorts, from recruitment through completion, to ensure Fellows have everything they need to meet their goals. The PM needs to have someone create and maintain the websites for each of the cohorts, collect and store classroom video recordings, and create quarterly reports. The PM may also create program newsletters as well as invitations and programs for special events. The PM will keep track of spending of grant funds to ensure spending is aligned with the budget plans. This will require that the PM understands the university procurement system and complies with university spending and record-keeping policies. The PM also oversees the work of the advisors with regular checks on the Year 2 Fellows' files to ensure Fellows are receiving the support they require. The PM at the lead university will also have the additional responsibility of compiling and summarizing the quarterly reports from all other partner university sites.

Assistant to the Program Manager/program administrator

A university may find that the job of the PM can be better managed with assistance. The primary responsibility of the Assistant to the PM or Program Administrator (PA) is to assist in day-to-day functioning related to the events and administration of the SEF program. This role is particularly helpful if the roles of PM and Budget Manager are combined or if the PM is assisting the PI in leading a significant research agenda. In particular, the PA handles any/all clerical work associated with the fellowship, such as copying, emailing reminders, setting up forms and surveys when needed, and ordering materials. The PA may also take care of communication with cohorts, including sending acceptance letters, invitations, information on meetings, etc., and may coordinate with other university sites about conferences and information that they might need about running the program. The PA also coordinates details of the yearly celebrations, such as organizing table and chair set-up, booking event spaces and catering, communicating with tech support, creating signage, and printing name tags. The PA may also keep track of contact lists for each cohort.

Office/budget/grant manager

Whether you are running this program from a large research institution or a small community college, there needs to be a group to establish and review contracts, budgets, payments, and review research proposals for ethics and human subject protection concerns. Contracts and grant establishment are generally handled through an Office of

Research and Sponsored Programs (ORSP – the grant office). All institutions must have some budgeting system for accountability to funding sponsors/agencies/government. The budgeting system may also be the payment system, or a separate accounts payable system may exist. Lastly, if any research is to be conducted in parallel to the administration of the program (and we strongly encourage this), there must be an Institutional Review Board (IRB) or ethics panel to ensure all research is conducted at the highest ethical standards and the rights of all human subjects/participants are protected. This becomes even more critical when working with potentially vulnerable groups such as minors and persons from historically marginalized groups.

Funding for a program like this can come from private companies, foundations, or non-profit agencies. An alternative would be to have the five local school districts fund major parts of the program through their professional learning allowances. Once the funding has been secured, the fiscal sponsor and the lead university sign a contract. The contract and budget are established with the assistance of the university's ORSP (grant office). Once the funder meets the budget obligations, the budget becomes part of the university's budget system. The key aspect is that a contract and budget must be in place for the work to begin within the program.

The SEF budget manager will be a university staff member proficient in the university budget system and will oversee the spending on the grant budget and all subcontracts. The budget manager will assist in getting the account for the grant in the system and will keep track of spending on a regular basis. The lead university budget manager will work with ORSP to finalize subcontracts and get partner universities budgets in place. The budget manager also will procure items at the request of the PI and PM (described later). The budget manager will ensure that the spending of the grant budget is in compliance with university policies.

> *As the person responsible for the administrative and financial components of the Wipro grant, I work with the leadership team and create the applications, surveys, communications, and agendas throughout the Wipro program. I coordinate logistics for all meetings with the leadership and the Wipro Fellows. I am the main contact for all correspondence with the Wipro Fellows. If applicable, I oversee the work that is being accomplished by the Wipro Fellows, which includes purchases and providing stipends/incentives for participants.*
>
> Colette Killian, Montclair State University, NJ

Lead Instructor(s)/Facilitators

Lead Instructor(s)/Facilitators should be identified to plan and implement professional learning for Fellows throughout the course of the SEF program. The Lead Instructors will create a learning arc for the monthly cohort meetings, which can range in topics from in-depth understanding of the science practices as laid out in the Next Generation Science Standards (NGSS) and the best practices of teaching science K-12. The Lead Instructor(s) needs to be familiar with the core components of the SEF program, such as the V-CCLS and H-CCLS models and protocols, and be able to convey these protocols to the Fellows. The Lead Instructor(s) will also collaborate with District Coordinators on current district priorities for science and will tie these priorities into the SEF program.

Doctoral students

Doctoral students in teacher education programs at the universities may support the program by assisting on tasks including handbook design, conference planning, and Fellow support. Doctoral students may also assist Lead Instructors with facilitation of professional learning and mentoring Fellows on their growth projects. Additionally, doctoral students may be involved in research projects with the PI and/or Lead Instructors.

The Wipro CA Program has strong leadership that works synergistically to run a coherent program. The key roles and responsibilities are outlined below:

Wipro SEF CA – Center to Support Excellence in Teaching

- *Principal Investigator (PI): Oversees the conceptual direction of the program.*
- *Program director: Oversees all parts of the program, leads the team in implementing all components, ensures high-quality professional learning, and ensures responsiveness to Fellows' and district needs.*
- *Financial administrator: Manages the overall budget, tracks spending, and ensures alignment with institutional requirements.*
- *Professional learning facilitators and coaches: Create and facilitate professional learning experiences and regularly meet with individual Fellows throughout their time in the program.*
- *Administrator: Handles logistics and event planning.*
- *Doctoral students: Assists with professional learning content and research organization.*
- *Graphic designer: Creates programs, brochures, and other media for major Wipro events.*

Tammy Wu Moriarty, Stanford University, CA

Year 2 GPS advisors

In Year 2, the Fellows implement their growth projects or GPS while meeting regularly with their assigned GPS advisor. Year 2 GPS advisors may or may not be affiliated with the university and each advisor may work with more than one Fellow. Advisors can be university faculty and staff as well as former SEF Fellows or DSCs (ideally from a district other than the Fellow's district). The responsibilities of the advisor are to first guide Fellows through the writing process of their plan by providing feedback, helping each Fellow focus his/her plan on reasonable goals, and then have regular advising sessions with their Fellows throughout the year. The advisor is responsible for meeting with the Fellows to check in with their progress and to suggest alternative plans and revised timelines if needed. Advisors will report back to the SEF leadership team and the DSC if there are any concerns about Fellows not meeting benchmarks. Each advisor should be someone with whom the Fellow can make a connection and who might have a particular strength in the areas concerning the Fellow's GPS plan. It is important that the advisor does not play a district supervisory position to the Fellow as that may impact the nature of their relationship.

The SEF leadership team

Within the first months of moving forward with the program (i.e., possibly receiving a grant) and after selecting the partner districts, it is important to develop a strong collaborative team. In addition to tactical administration of the program, it is important to have a group that can meet and review the program from a more strategic perspective to ensure that the overall vision of the program is coming to fruition. The SEF leadership team at each site is made up of the university PI, the PM, Assistant to the PM/Administrator, Lead Instructor(s), and the DSCs. During the first year of program implementation, when all the key players are getting to know one another, it is important to meet frequently as a leadership team. The leadership team sets the course for the program, develops the monthly meeting agendas, creates the learning arc, and discusses current challenges and successes of the program. They also plan all special events and put a recruitment plan in place, keeping in mind the goals for each district for the upcoming cohort. Leadership team meetings are a chance to reflect on program implementation and to determine any changes that need to be made. Once the program is fully established, the frequency of the leadership team meetings can decrease.

With the many intricacies of the SEF Program, the leadership team must be a strong and collaborative team. One way to strengthen the team is to plan a retreat in a location outside of the university site so members of the leadership team can learn more about each other, the core tenants of the SEF program, and the roles they will play as the program is implemented. During the retreat, you may consider scheduling all future leadership meetings, cross-site meetings, as well as monthly cohort meetings for the upcoming year. You may also consider setting norms and establishing problem-solving protocols so that challenges to reaching the SEF goals can be met with decisions about possible solutions.

External evaluator

It is always valuable to have an external program evaluator to collect data and report key findings so that the SEF program can continue to improve. The external evaluator may attend some monthly IHE meetings and conducts regular surveys distributed to all Fellows and DSCs as well as interviews with Fellows, school administrators, DSCs, and SEF program staff. It is crucial for the sites to work collaboratively and coordinate with the external evaluator as to avoid overwhelming the Fellows, DSCs, or other participants with surveys. In the worst case scenario, the external evaluator sends out a survey and the site leaders send out a second survey at almost the same time. This results in confusion and decreases the response rate across both surveys. In the best case scenario, surveys and research from the external evaluator and the local site leaders are coordinated (and perhaps even combined to avoid redundancies) to limit possible survey fatigue.

Key SEF tasks

In the previous section, we explained the tactical or day-to-day roles and responsibilities of key personnel needed to implement the SEF. The following section will describe the key tasks for effective program implementation.

SEF meetings

The meetings throughout the years of the program are the glue that holds all partici-pants together. In addition to the meetings that the Fellows have in their CCLS groups, there are meetings of the leadership team, the DSCs and, in our present situation, cross-site leadership meetings of the seven university sites. Appendix 6A outlines the purpose of these meetings, the attendees, and their frequency.

Recruitment and selection of partner school districts

At the onset of the SEF program, recruiting and selection of local partner school districts is one of the first tasks. If the university has a significant teacher education and prepara-tion mission, then it is likely that there are already long-established relationships between the university and local school districts. These school districts are ideal candidates for the SEF. There may also be local districts with whom the university would like to establish more significant relationships. These are also ideal candidate districts for the SEF.

When considering a geographic region from which to identify eligible districts, it's im-portant to consider distance and traveling time to the university where much of the pro-fessional development will take place. To some extent, this has been ameliorated by the development of robust and viable video meeting technology such as Zoom or Google Meet. Still, some of the functions of this program are best held in-person. The SEF leadership team may keep in mind other factors such as percentage of high needs or low-income students when recruiting school districts from the determined geographic region. Ideally, the characteristics of the chosen school districts should be in-line with the overall mission of the university. Below, we will outline the key steps in recruiting partner school districts to be a part of the SEF program.

1 **Identify potential districts**

Once the geographic region has been determined, most state Departments of Edu-cation have databases that contain the necessary information needed by the leader-ship team: the name of the district, the superintendent of the district, the student population, and number of schools. From this database, a list of potential partner districts and schools can be created. Begin communications with these districts with a letter announcing the SEF program opportunity along with the program bro-chure. The letter should provide information details of upcoming information ses-sions that will be held for interested districts. It is likely that the SEF leadership team has already formed relationships with science leaders in these districts and it would be appropriate to personally reach out to these districts.

2 **Create district applications/request for proposal**

An application or RFP is created for the purpose of learning about the district, the science programs already in place in the district, and its commitment to K-12 sci-ence education. Several open-response questions should be asked to determine the structure and ability of the district to support the SEF program. The application should also include a "Partner Contributions and Commitment" form that the su-perintendent signs to ensure the district is capable and fully aware of the required program components and necessary teacher support. See sample questions and agreement in Appendix 6B.

A Roadmap for Transformative Science Teacher Leadership

3   **Hold information sessions and support applicants**

The PI and PM should plan, schedule, and hold information sessions at the university or other central location geared toward educating district representatives about the SEF program and its benefits to the district and its K-12 teachers of science. At these sessions, the core components of the SEF program and its pillars should be discussed in detail. A careful review of the requirements noted in Chapter 3, specifically that teacher-Fellows are to record classroom videos and post them to a secure site and share them with Fellows across the program using a specific "tuning" protocol, should be noted as non-negotiable. (It is important that districts agree to allow this and do not find this requirement a burden, as this is a central component of the work of the teachers in the first year of the program.) In parallel with planning the information sessions, the PM should distribute the application or RFP and provide time during the information session to review the requirements of the application. It is important to allow for a question-and-answer period and collect participant information so the PM can follow up with those who were interested in becoming an SEF partner district. Alternatively, information sessions can be held as a webinar using an application such as Zoom or Google Meet.

During the application period, it is important that the PM is available to answer questions and provide support to the district representative who is leading the effort to submit the application. The form is an important component for capturing the most timely and important information about a district.

4   **Select partner districts**

Once completed applications or responses to the RFP have been submitted, the SEF IHE leadership team, in addition to other trusted colleagues, evaluates the applications using the agreed-upon scoring rubric. Depending on the number and strength of the applicants, the SEF IHE leadership team can decide if meeting with key district personnel is necessary before selecting the final five partner districts. The SEF leadership team may also decide because of district factors like size and geographic span if more than or fewer than five districts should be selected. As we've mentioned, a single district can implement the SEF program. However, there is much gained by the interactions of teachers from other school districts.

5   **Announce awards to the districts and press**

Once the final decision has been made about the partner districts, send a letter to the superintendent with a copy to the DSC offering the partnership to the district. In the letter, ask to meet with the superintendent (or designee) and the DSC to finalize the partnership and answer any questions. The PI and/or PM arranges, with the superintendent of the district, a face-to-face meeting. At this meeting, the PI or PM explains the goals of the program and the importance of recruiting teachers from all grade spans and content areas, preferably with a representative sample of teachers from elementary, middle, and high schools. The first meeting with the district administration sets the tone for the relationship between the SEF leadership and the district partnership. It is important that the benefits, key components, and structure of the program are communicated well, and the administration feels that the SEF leadership is aligned with the science initiatives of the district. The letter should also invite the DSC to the first meeting of the partner districts with the SEF leadership team.

Once all the districts have confirmed their partnership in the SEF Program, collaborate with the university communications department to send out a press release to the regional newspapers. We recommend sending a press release to each of the districts, as many districts have ways to announce new programs through their own district website. Having this new partnership announced at a school board meeting will generate enthusiasm for the anticipated positive changes in the school district.

Recruitment and selection of Fellows

Once the university has recruited and selected the school district partners, the next task is the recruitment of the first cohort of SEF Fellows. In coordination with the DSCs, plan a recruitment timeline and process. Ideally, recruitment of the Fellows occurs from December through March or April. Recruitment, as a core tenant of the SEF program, is a collaborative effort between the university and the district partners. Below is an outline for the steps involved with the recruitment of teacher-Fellows.

1   **Eligibility**
    In general, eligible teachers are those teachers in grades K-12 who have taught science for at least three years and been employed by their current district for at least two years. High school and middle school teachers must hold a valid license to teach a science subject or general science. Elementary teachers must hold an elementary license as a classroom teacher. Teachers also must have the support of their school administrator to apply to the SEF program as well as receive consent to record video of their classroom practice. Furthermore, teachers who become Fellows are committing to remain in the classroom for the entire term of their fellowship. Teachers should be aware that SEF is an intensive program that will require an average of 125 hours of their time each year (outside their contractual school hours), and applicants should carefully consider their teaching and professional obligations before applying.

2   **Create Fellow applications and scoring rubrics**
    The application process can be set up electronically or paper-based, but requires teacher candidates to respond thoughtfully to several open-ended questions. The application should ask for enough information about the teacher to get a sense of their classroom experience, their leadership experience at their current and previous schools, and their content and pedagogical depth. Teachers should be able to describe past professional leadership experiences and must be able to demonstrate how they have been successful in the classroom. SEF sites can determine what documentation they will require applicants to submit such as a current resume or teaching certificate. The IHE should be prepared to provide an application that the partner district can adopt or modify to suit its own needs. All applicants should be required to sign, along with their principal and DSC, a fellowship agreement that acknowledges the primary responsibilities of the Fellow, the principal, the DSC and the SEF program during the fellowship years.

3   **Distribute Fellow applications to partner school districts**
    The site leadership team will send SEF applications to teachers in the partner school districts with a brief introductory email. Some universities may add an "intent to

apply" form and only distribute applications to those who have completed this form. The benefit of using this method is that the PM can keep track of how many teachers from each district and each grade level are intending on applying, which can help with meeting recruitment goals from each district. Others have sent the applications out to partner school districts more broadly. The SEF leadership team can also send applications to the DSCs to distribute. The applicants may have questions or concerns, so it will be important to include contact information for the SEF PM. As the program matures, we have found that alumnae SEF Fellows are the best recruiters because they can talk directly about their experiences in the program and teachers are more likely to listen to their peers.

4   **Meeting with district and school leadership**

During teacher recruitment for the program, it is vital that administrative support is in place. If the administration understands the vertical nature of the SEF program, the need for teachers from the K-12 spectrum, and why the SEF program is beneficial to the district, they will be likely to ask principals to nominate and encourage teachers to apply. Additionally, if school leaders would like to know more about the program, it is important to make the time and space to meet with principals or assistant principals to describe the programs benefits to Fellows, the school, and the district.

5   **District Science Coordinator assistance with recruitment**

With the support of the district administration, the next step to take is to have the DSCs assist with all recruitment efforts for their respective districts. This may include making announcements about the fellowship program to all the science teachers in the district, including all elementary teachers. The announcement should have information about the SEF program as well as all relevant contact information and links to program applications and information listed on university websites. DSCs have knowledge and relationships with teachers in their districts and can help with the recruitment process by giving encouragement to teachers who would be a good fit for the SEF Program. They also have a better understanding of their district landscape, including which school principals would be most supportive of their teachers becoming Fellows in the SEF Program.

6   **Hold university and district-based information sessions**

Informational meetings are scheduled at each district and facilitated by the DSC, the PM, and in subsequent years, a former Fellow from a previous SEF cohort. The PM will present the program and answer questions from interested teachers. Information can be shared through a presentation as well as with printed brochures. Given the nature of teachers' schedules, it is recommended to have at least two information sessions – one during an evening and one immediately after school. Either or both can be held in-person or virtually. The presentation should include links to the SEF university website and/or form for Fellows to complete stating their intent to apply.

7   **Read and score applications**

After all applications for the SEF have been submitted, reviewers consisting of SEF faculty and staff and DSCs will use a scoring rubric to review applications. The SEF leadership team at each university will discuss each applicant. You may wish to limit the DSCs formal input about Fellows from their own districts. DSCs should be able to provide their professional opinion of teacher candidates from their districts in much the same way as they would in a letter of support to other programs to which their teachers apply. However, it will be important for each university to agree on a process that eliminates or limits selection bias.

8 **Make final selection of Teacher-Fellows to ensure program targets**

The goal of the SEF program is to have enough Fellows, over time, at each district across grade bands to make a significant impact on district science initiatives. For instance, this could be something like having 12 Fellows, spread among the three grade spans, in each of the five partner districts. Keeping this target recruitment in mind is important as there will be variations in numbers of applicants and grade levels of applicants during each year of recruitment. Within each cohort, however, there needs to be representation from the three grade spans (elementary, middle, and high school) to support the vertical groups described in the V-CCLS section, which is a core component of the SEF program.

A matrix such as the one in Table 6.2 can be created to include all applicants from all five districts. This matrix is useful during the selection process.

The SEF leadership team must carefully balance the number of high school, middle school, and elementary science teachers admitted to the program and ensure there is balanced representation of science teachers from the five partner districts and from the physics, chemistry, life, and Earth sciences. It is this balancing across the five districts that justifies the centralization of Fellow applicant evaluation and

Table 6.2 Sample applicant selection matrix

		Biology	Chemistry	Earth/Environmental Science	Physics/ Physical
District 1	K-2	Applicant 1 C2 (K)	Applicant 9 C2 (K-5)	Applicant 17 C2 (K-5)	Applicant 21 C2 (K-5)
District 2		Applicant 2 C2 (K)	Applicant 10 C2 (K-5)		
District 3	Grades 3-5	Applicant 3 C2 (G3)	Applicant 11 C2 (G4&5)	Applicant 18 C2 (G4 & 5)	Applicant 22 C2 (G5)
District 4	Grades 6-8	Applicant 4 C2 (G7) Gen Sci	Applicant 12 C2 (G6) Gen Sci	Applicant 19 C2 (G6)	Applicant 23 C2 (G8) Gen Sci
District 5		Applicant 5 C2 (G 6-8) Gen Sci	Applicant 13 C2 (G8) Gen Sci		
	High School	Applicant 6 C2 (HS Bio)	Applicant 14 C2 (HS Chem)	Applicant 20 C2 (HS Earth & Env Sci)	Applicant 24 C2 (Physics– ELL)
		Applicant 7 C2 (HS Bio)	Applicant 15 C2 (Chem)		Applicant 25 C2 (Physics)
		Applicant 8 C2 (HS Bio)	Applicant 16 C2 (HS Chem)		

selection. When selecting a new cohort of teachers and placing teachers in vertical teams, consideration should be given to the distribution of teachers across districts. If a teacher teaches two subject areas, such as chemistry and biology, there is some flexibility in where that teacher can be placed in the V-CCLS matrix. There is also flexibility where K-8 teachers are placed in subject groups since typically K-8 teachers teach varied science content. A total of about 20 K-12 teachers should be selected to be SEF Fellows for each cohort.

In subsequent years and cohorts, to ensure parity across districts and subject areas, it will be important to look at the current enrollment of Cohort 1 when selecting Cohort 2, or keep in mind the make-up of Cohorts 1 and 2 when selecting Cohort 3. This balancing element is added to the process after Cohort 1 Fellows are selected. An idealized distribution of Fellows across cohorts is provided in Table 6.3.

9   **Letters of congratulations and regret**

Once the final decisions have been made about the Fellows for each cohort, the university SEF leadership team sends letters of congratulations to those who were selected. In this letter, Fellows are asked to confirm acceptance into the SEF Program and to "save-the-date" for an induction ceremony and celebration. Letters of regret are also sent to those who were not selected. Strong candidates who were not chosen due to the number of applicants who applied, or due to the number of limited spaces in each district, may be given a spot in the next cohort and will not need to re-apply the following year. Additionally, the university SEF leadership team sends letters to each of the teacher's principals and the district superintendent informing them of the award to their teacher(s).

Communications plan

In this age of media and branding, it is important to advertise and announce the award of the SEF program. The communications strategy is multifaceted. It should include both traditional forms of communication as well as social media and be internal and external to the university. Listed next are various forms of communication needed by the leadership team at each university.

1   **Press releases**

The press release should mention key people at the university and the funder who were involved in the negotiations of the contract. It should name the funding amount and the number of years for which the project will take place. The press release should note the partnership between the IHE and the local partner school districts and highlight the major tenets of the SEF program including that the SEF:

○ Supports three cohorts of 20 science Teacher-Fellows (a total of 60 teachers) from five local districts during a two-year fellowship that focuses on leadership in science education.

○ Promotes curriculum articulation within districts through vertical and horizontal team lesson study across districts.

Table 6.3 Idealized balanced Fellow distribution across three cohorts of SEF

Ideal Configuration of Wipro Fellows Cohorts 1 & 2 & 3 – 12 Fellows from each district distributed among grade levels				
	Biology	*Chemistry*	*Physics*	*Earth Science*
(K-5) Elementary School Teachers/ Fellows	Fellow 1 Cohort 1	Fellow 2 Cohort 1	Fellow 5 Cohort 2	Fellow 9 Cohort 3
	Fellow 9 Cohort 3	Fellow 1 Cohort 1	Fellow 2 Cohort 1	Fellow 5 Cohort 2
	Fellow 5 Cohort 2	Fellow 9 Cohort 3	Fellow 1 Cohort 1	Fellow 2 Cohort 1
	Fellow 2 Cohort 1	Fellow 5 Cohort 2	Fellow 9 Cohort 3	Fellow 1 Cohort 1
	Fellow 11 Cohort 3	Fellow 12 Cohort 3	Fellow 5 Cohort 2	Fellow 9 Cohort 3
(6-8) Middle School Teachers/Fellows	Fellow 6 Cohort 2	Fellow 7 Cohort 2	Fellow 3 Cohort 1	Fellow 10 Cohort 3
	Fellow 10 Cohort 3	Fellow 7 Cohort 2	Fellow 6 Cohort 1	Fellow 3 Cohort 1
	Fellow 3 Cohort 1	Fellow 10 Cohort 3	Fellow 7 Cohort 2	Fellow 6 Cohort 2
	Fellow 6 Cohort 2	Fellow 3 Cohort 1	Fellow 10 Cohort 3	Fellow 7 Cohort 2
	Fellow 7 Cohort 2	Fellow 6 Cohort 2	Fellow 3 Cohort 1	Fellow 10 Cohort 3
(9-12) High School Teachers/Fellows	Fellow 11 Cohort 3	Fellow 12 Cohort 3	Fellow 8 Cohort 2	Fellow 4 Cohort 1
	Fellow 4 Cohort 1	Fellow 11 Cohort 3	Fellow 12 Cohort 3	Fellow 8 Cohort 2
	Fellow 8 Cohort 2	Fellow 4 Cohort 1	Fellow 11 Cohort 3	Fellow 12 Cohort 3
	Fellow 12 Cohort 3	Fellow 8 Cohort 2	Fellow 4 Cohort 1	Fellow 11 Cohort 3
	Fellow 1 Cohort 1	Fellow 2 Cohort 1	Fellow 8 Cohort 2	Fellow 4 Cohort 1

Color key	**District 1**	**District 2**	**District 3**	**District 4**	**District 5**

- ○ Promotes research-based pedagogical best-practices.
- ○ Builds teacher leadership capacity by having Fellows focus on district-wide science initiative and leading professional development in their own districts.

It's also important to get the word out about the SEF award to the broader community for the purpose of giving credit to the funder. In our case, our sponsor has an employee base of more than 170,000 people spread over 50+ countries. Funding a program like this serves a dual purpose: It shows the commitment the sponsor has to the communities it serves and it helps the sponsor meet its goal of attracting and retaining the best STEM talent from a globally diverse talent pool, which in turn helps to better serve customers and provides a strong competitive edge in the global marketplace. The logic model for STEM professional development programs like the SEF is that better trained K-12 science teachers will ignite more interest in STEM college and career paths for students.

Upon selection of Fellows, work once again with the university's communication department to send out a press release in each of the five communities announcing the teachers in their district who have been selected. A template is created for all releases. The announcement should name the Fellows and the district and schools in which they teach. It should state the monetary amount of the Fellowship, the goals of the program and that the funding is from the sponsor (if applicable). Once the template has been created, the information is completed for each district. Using the same template as above, the PM can create press releases that can go to the hometown newspapers of each of the Fellows. A completed press release can be sent to each Fellow to send to their Hometown or district/school newspaper.

2 **Website**

The website for the SEF program will be placed within the larger university's website and will take on the branding of the university in conjunction with the use of the SEF logo. The website for the SEF program should describe the key components of the program, list the partner districts, and explain the Fellow application process. The website should have contact information and list key program elements. Depending on the university, this website can also act as a portal for document storage for the Fellows. The website should also be current with dates in regards to the recruitment period and application deadlines for each cohort.

Links to several examples of university websites are listed here:
https://www.mercy.edu/academics/center-stem-education/wipro-science-education
https://physics.missouri.edu/wipro-fellowship
https://wiprostemprogram.com/

3 **Printed/Online brochures**

Each university SEF leadership team should design and develop brochures for two different audiences: the district administration and the teacher recruit. The brochures can be nearly identical with the difference being the benefits of the program to that audience. The brochures will be used to recruit both the districts and the Teacher-Fellows once the districts have been selected. Included in the brochure should be an outline of benefits. See Appendix 6D for an example list of benefits. Once all the print materials for the recruitment process have been created, it is important to put these artifacts on the SEF website.

Celebration and induction of each cohort

Much like application to college or graduate school, the Fellows have persevered through a challenging process and have made a commitment that will have a significant impact on their personal and professional lives for the next few years. This fact should not be lost in the desire to get to the good work of the SEF. Along with the press releases and announcements noted earlier, Fellows are invited to an induction celebration of their acceptance into the SEF program. This induction ceremony could occur in the late spring or early fall. During the induction ceremony, all Fellows within a cohort will meet each other for the first time. The feel of the celebration should be upbeat and supported with the appropriate décor. For instance, you may consider presenting the Fellows with corsages or boutonnieres to distinguish them from other guests for the Fellows to know which attendees are going to be in their cohort. This distinction will also be a way to foster conversations. The program of the celebration should include time for mingling and eating hors-d'oeuvres, and listening to invited speakers, including an address from the PI or PM of the SEF program. You may consider inviting dignitaries from the university and funding sponsor as well. A professional photographer is also important to take individual portrait photographs of the new Fellows as well as group photographs of the Fellows with their district groups and respective DSCs. It is also important to take a photo of the newly inducted cohort of Fellows as a whole group that can be used for press releases and the website. An example induction ceremony checklist is provided in Appendix 6E along with an example induction ceremony program.

Professional learning sessions

One of the most important aspects of the SEF program is the monthly professional learning session. It is essential for the university leadership team to create a meaningful learning arc that strengthens science teaching and learning and aligns with partner district science goals. The professional learning facilitation team may include professors, doctoral students, and/or program staff with expertise in different areas. These professional learning sessions can be in-person or virtual, but we highly recommend as many in-person sessions as possible that make sense for your context. The cadence of these sessions is also important to consider when thinking about district calendars and meetings. Planning meeting times that do not conflict with school and district events is essential in the ability of Fellows to participate in the learning sessions.

Reporting (if appropriate)

The final major responsibility of the IHE in the SEF program is reporting. Each quarter (December 1, March 1, June 1 and September 1), an operations and fiscal report is due to the sponsor detailing program successes, challenges, and events of the previous quarter. The report will also include important upcoming dates, next steps that will be taken to implement the program, and a brief interim report from the program evaluator. An example template sent to each site is shown in Appendix 6F.

Conclusion

In presenting the details of selection of districts, recruitment of Fellows, and other logistics, we hope that we have not pushed to the background the value of the SEF program for the IHE, the district, the Fellows, and the DSC.

Universities have a unique opportunity to enhance their impact and fulfill their educational mission by collaborating with local school districts on science education initiatives like the SEF. This partnership offers numerous benefits that extend beyond the immediate educational outcomes for K-12 students. Here are several compelling reasons why universities should consider engaging in the SEF program:

1 **Enhanced research opportunities** – Partnering with local school districts provides universities with access to a real-world laboratory where innovative approaches to fostering teacher leadership and its impact on students can be tested and refined.
2 **Community engagement and outreach** – By engaging with local school districts, IHEs foster a positive relationship with the surrounding community, enhancing the IHE's reputation as a proactive and engaged institution.
3 **Student recruitment and retention** – Collaborative initiatives in science education can serve as a recruitment tool for universities. Through the SEF, teachers form meaningful relationships with IHE faculty and staff. This personal relationship with a local IHE leads to favorable views of the IHE for their graduating high school students.
4 **Grant and funding opportunities** – Collaborative projects like SEF between universities and local school districts are eligible for grant funding from governmental agencies, private foundations, and corporate sponsors. Securing grants for this program can lead to additional support from similar agencies.
5 **Real-world impact and social responsibility** – Universities have a social responsibility to contribute to the betterment of society. By partnering with local school districts in the SEF, IHEs can directly influence the quality of education, address educational disparities, and promote scientific literacy among the next generation.
6 **Interdisciplinary collaboration** – Science education initiatives like the SEF can benefit from expertise from multiple disciplines, including education, psychology, technology, and the natural sciences. Such projects foster interdisciplinary collaboration within the university, encouraging faculty and students from different departments to work together toward a common goal.

If you began reading this chapter so invigorated by the prospect of the benefits of this program, we hope that we have given you the final pieces to help you along your journey. If you began this chapter a little apprehensive about the commitment of the program and the amount of effort to run it, we hope that we have both given you a candid description of the program and shared with you some resources that will enable you to begin. We have lived this program for more than a decade and although we have benefitted from a grantor and developed as a lead university with six strong IHE partner sites, we are confident that a dedicated educator can rally some colleagues within their district or nearby, and develop and implement their own local deployment of this program. We invite you to join us in this work. We have even more resources than have been provided so far. We hope it is the beginning of a much more meaningful and sustained journey to champion district transformation through teacher leadership.

Appendices

Appendix 6A: Sample summary of meetings for key personnel in the SEF

SEF leadership meetings (non-Fellow) that occur in all years		
Meeting	*Who attends*	*Frequency of meetings*
Leadership team meetings (each site)	District Coordinators and SEF Staff	5 times per year* (Try to coordinate these with the monthly cohort meetings)
District Coordinator PLC meetings	District Science Coordinators only	4 times per year*
Cross-site leadership team phone conferences	Leadership team from each site	Monthly
Cross-site leadership retreat	District Coordinators and SEF staff from all university sites	1 time at the end of Year 1 for each cohort (once a year for the first 3 years). (Try to coordinate this with the Teacher Leadership Conference.)
Fellow events	See Fellows meeting chart for details	See Fellows meeting chart for details

Appendix 6B: Sample district applications to join SEF

Science Education Fellowship: District Application

Congratulations! Your district has been selected as a finalist for the Science Education Fellowship. We are looking to partner with four or five local school districts over the next five years to support 60 teachers. To help determine which districts will be selected, we will proceed with a two-step application process. The first part is to complete the written application and the second part is to have the project leadership team meet with the superintendent and you in a one-on-one meeting. **Applications are due on November xx, xxxx And can be sent electronically to [project manager].** Please let us know if you have any questions as you complete the application.

Applicant Information

Name of District: _____

Name of Applicant: _____

Title of Applicant: _____

Phone Number: _____ Alternate Phone Number: _____

Preferred E-mail Address: _____

Alternate E-mail Address (if applicable): _____

A Roadmap for Transformative Science Teacher Leadership

District Information

Name of Superintendent: _____

Name of District Science Coordinator: _____ Grade Levels Supported: _____

Phone Number: Alternate Phone Number: _____

Preferred E-mail Address: _____

Alternate E-mail Address (if applicable): _____

Science Department's Mailing Address: _____

City: _____ State: _____ ZIP Code: _____

District Science Department Data

Does your district have common curricula used throughout
the district?   Yes   No

Please describe: _____

Does your district use a module-based curriculum at the
elementary grades?   Yes   No

Please describe: _____

Does your district use a module-based curriculum at the
middle grades?   Yes   No

Please describe: _____

Does your high school(s) offer AP courses?

Please describe: _____

Does your district have science coaches at the district level? If yes,
how many?   Yes   No

How many teachers of science are in your district's elementary schools? _____

How many middle school science teachers are in your district? _____

How many high school science teachers are in your district? _____

Does your district use science specialists, classroom teachers, or a mixed model at the

elementary grade level? _____

Who are the main providers of science professional development to the teachers in your district? _____

Do you have a cohort of science teacher-leaders that presently exist in your district? If so, how many? _____

District Level Data

Number of Elementary Schools: _____ Number of Elementary Students: _____

Number of Middle Schools: _____ Number of Middle School Students: _____

Number of High Schools: _____ Number of High School Students: _____

Student Diversity: _____% Hispanic _____% Black _____% White _____
% Asian _____% Other/Multiracial

Student ELL Population: _____% ELL or LEP Student SES Population: _____
% Free- or Reduced-Lunch

Student Special Needs Population: _____% of students are enrolled in a special needs class

What percent of your district's operating budget is devoted to professional development?

Open Response Questions

Please answer the following questions to help us with the selection process. We are looking for districts that will be a good partner for this program. Please limit your responses to between 4 and 8 pages single spaced. If you are answering them on a separate sheet of paper, please clearly label each question.

1 **How does your district's partnership in the SEF program align with your vision for science in your school district?**
2 **Please describe your science program at the elementary school, middle school, and high school levels (including AP courses and level of participation). Include budgets devoted to science at each level.**
3 **How does the science department currently support the professional development of science teachers in your district? Within your answer, please describe the structure of the science department.**

4   The Science Education Fellowship is intended to recognize committed teachers of science who are poised and ready to learn, to collaborate, and to be reflective on their classroom practice.

    a   What do you think are key factors in identifying teachers who are ready for this opportunity?

    b   What will be your district's recruitment strategy to make sure teachers are applying?

5   Through participation in this program, what changes do you envision will occur over the next five years in your school district that most likely would not happen otherwise?

6   What are your strengths as a school district in the development of its teacher-leaders? What is one weakness or barrier you have experienced in the past when trying to develop teacher-leaders?

7   In Year 1 of the fellowship, each teacher will work with all Fellows along a number of initiatives (e.g., vertical articulation of curriculum and horizontal articulation). Each teacher will also work with the district. In Year 2 of the fellowship, each teacher will be pursuing a personal professional development plan which will support both district goals and individual enrichment. Describe how you will structure this time for the district meetings and how you can best utilize these Fellows.

8   What is one thing you would like the committee to know as districts are selected to partner for the SEF program?

Appendix 6C: List of benefits of the SEF program

The benefits to the district in supporting teachers in the SEF program are that the district gains a core of K-12 teachers of science who:

- Have greater knowledge and understanding of researched-based science teaching practices.
- Have had experience learning and using tools of lesson observation.
- Are more reflective practitioners.
- Have led professional development in science teaching and learning.
- Know how to plan and implement an action plan related to promote a district initiative.
- Have stronger leadership skills.

The benefits to the District Science Coordinators are that they

- Form relationships with DSCs from the other partner districts.
- Learn more about their Fellows by working with them in the context of the SEF program in the university setting.
- Gain a group of science teachers on whom they can call when new district initiatives are being discussed and implemented.
- Build a relationship with IHE professionals.

The benefits to the Science Education Fellows are that opportunities are provided to gain skills in the three pillars of SEF:

Leadership

- ○ Develop relationships with teachers across districts, content areas, and grade levels.
- ○ Plan and lead professional development in science education.

Adult Learning

- ○ Dive deeply into specific science content and science and engineering practices (e.g., NGSS) with their SEF colleagues.
- ○ Develop and implement an individual growth plan with ties to a district science initiative.

Reflective Practice

- ○ Learn and use the Collaborative Coaching and Learning Science Protocol for observing and reflecting upon their own and others' classroom videotapes.

Appendix 6D: Example induction ceremony checklist and program

☑ Induction Ceremony Checklist	Delete

0% ▬▬▬▬▬▬▬▬▬▬▬▬▬▬▬▬▬▬▬▬▬▬▬▬▬▬▬

- [ ] Adequate budget allocation. (There is no supplementary budget for the induction ceremony.)
- [ ] Set date for ceremony.
- [ ] Select venue.
- [ ] Decide on catering for event. It can be as simple as coffee, soft drinks and snacks. Some have had cocktail hours prior to the ceremony.
- [ ] Invite Wipro representatives and invited speakers.
- [ ] Secure Photographer for event.
- [ ] Poll Fellows for names and addresses of people they would like to invite.
- [ ] Design Invitation
- [ ] Finalize Invitation list - send out invitations with RSVP information
- [ ] Confirm guest list.
- [ ] Reconfirm catering with exact numbers of people.
- [ ] Order flowers for each fellow (or other recognition). A colored name tag will work. You want everybody to be able to pick out the Fellows during coffee in order to congratulate them.
- [ ] Program which should recognize all districts and have the names of all Fellows being inducted. See sample from UMass
- [ ] Need photos and bios of all fellows - NY did a PPT scroll during coctail hour portion
- [ ] NV, NJ and Boston did corsages for Fellows. UNT did metal nametags
- [ ] Ask President and or dean to speak
- [ ] Ask Arthur to speak
- [ ] Invite Wipro Rep to join/speak

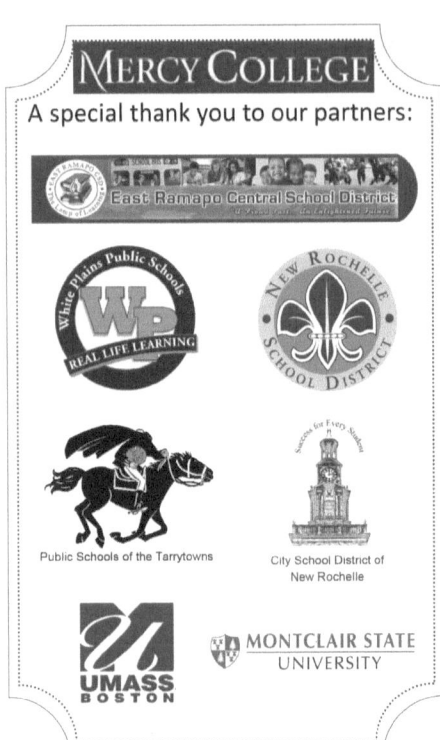

MERCY COLLEGE

A special thank you to our partners:

East Ramapo Central School District

White Plains Public Schools — WP REAL LIFE LEARNING

NEW ROCHELLE SCHOOL DISTRICT

Public Schools of the Tarrytowns

City School District of New Rochelle

UMASS BOSTON

MONTCLAIR STATE UNIVERSITY

WIPRO
Applying Thought

ScienceEducation Fellowship

Induction Celebration

Honoring Greater New York
Cohort One Fellows

June 16, 2014

MERCY COLLEGE

Congratulations to the 2014-2016 Wipro SEF Fellows!

Abbey Gilligan, *Public Schools of the Tarrytowns*

Ann Marie Ambrosino, *City School District of New Rochelle*

Carmen King, *White Plains Public Schools*

Charles Sincerbeaux, *White Plains Public Schools*

Claudia Gianserra, *City School District of New Rochelle*

Duane Stilwell, *East Ramapo Central School District*

Elizabeth Zahn, *City School District of New Rochelle*

Enrique Tovar, *Port Chester Public Schools*

Erika Sprauer, *East Ramapo Central School District*

Jannethe Pardo, *East Ramapo Central School District*

Karen Lent, *East Ramapo Central School District*

Karla Purcell, *Port Chester Public Schools*

Katherine Gentile, *Port Chester Public Schools*

Leana Peltier, *Public Schools of the Tarrytowns*

Liz Weiner, *East Ramapo Central School District*

Magali Dupuy, *East Ramapo Central School District*

Maureen Nimphius, *White Plains Public Schools*

Mayerlin Strippoli, *Public Schools of the Tarrytowns*

Sissi Johnson, *White Plains Public Schools*

Vincent Dougherty, *White Plains Public Schools*

Program

Photos

Head shots taken of all fellows; group fellow photo

Remarks

Meghan Marrero
GNY Wipro Co-director

Arthur Eisenkraft
Boston Wipro SEF Director

Alfred Posamentier
Dean, School of Education

President Timothy Hall
Mercy College

Francisca Godinho
Wipro Corporation

Mary Hall
MSU Wipro Fellow, Paramus Public Schools

Introduction of Fellows

by District Coordinators

Jason Choi, *Public Schools of the Tarrytowns*

Marselle Heywood, *City School District of New Rochelle*

Andrea Coddett, *East Ramapo Central School District*

Jessica Floridia, *White Plains Public Schools*

Elsy Zizolfo, *Port Chester Public Schools*

Closing

Amanda M. Gunning
GNY Wipro Co-director

A Roadmap for Transformative Science Teacher Leadership

Appendix 6E: Example template for quarterly report to funder

Institution	
Report Author	

Instructions: Please provide the information for each item. If you do not have a response for the item, please indicate the reason why this information does not pertain to your site. Please write your responses in the third person using Cambria 12 font, a line spacing of 1.15, and black ink. Please insert photos when appropriate and *caption each photo*.

Please submit your form with the following name:

Year 2 June date your state. Doc

1   Fellows' meetings (Cohort 2) (What has been the focus of your Fellows' meetings during this quarter March–June? Please include up to 2 captioned photos.)

    a   At the IHE?
    b   Within districts (DSCs and Fellows)
    c   At the informal science partners

2   Leadership meetings (What has been the focus of leadership meetings with the IHE, ISI, and DSCs)

3   End of the year host conference (Please describe the event: who was in attendance, highlights, successes, and challenges in the planning of the event. Please include photos with captions (limit 4) and attach the event program.)

Site Location (State)	Date of Conference	Conference Location

3a   H-CCLS Presentations (Cohort 2) (Include no more than 2 photos with captions.)

Team Name (include grade span)	Team Members	Science/Engineering Practice	Presentation Title

3b Conference Reflections

Site leadership reflections (What went well with the conference? What would you change/improve for next time?)

- H-CCLS presentations
- Keynote speaker
- Poster session
- Other

Reflections on the CCLS Teams and Presentations (Include any reflections you may have collected from the Fellows about the H-CCLS team experience and/or the presentations.)

Visitors from other SEF sites (What was the value of visitors from other sites at the conference?)

4   **Planning for Cohort 2 GPS** (Please describe your site's plan for how the Fellows will write their GPS.)

5   **GPS Poster Session (Cohort 1)** (Please describe the event: who was in attendance, highlights, successes, and challenges in the planning of the event. Are there things you would do differently next time? Please include photos (limit 4) and the event program.)

Site Location (state)	Date of Poster Session	Number in Attendance

a.)

Fellow's Name	Title of Poster	GPS Description

6   Recruitment and Induction of Cohort 3
6a Recruitment efforts (Please summarize the successes and challenges of your sites recruitment efforts) (250 words)

Applicants/District

District Name	Number of Applicants

Applicants by grade level

Grade level	Number of Applicants
K-5	
6-8	
9-12	

Please provide a matrix of Fellows by district and discipline (A sample is shown below)

Ideal Configuration of Wipro Fellows Cohorts 1 & 2 & 3 – 12 Fellows from each district distributed among grade levels				
	Biology	Chemistry	Physics	Earth Science
(K-5) Elementary School Teachers/ Fellows	Fellow 1 Cohort 1	Fellow 2 Cohort 1	Fellow 5 Cohort 2	Fellow 9 Cohort 3
	Fellow 9 Cohort 3	Fellow 1 Cohort 1	Fellow 2 Cohort 1	Fellow 5 Cohort 2
	Fellow 5 Cohort 2	Fellow 9 Cohort 3	Fellow 1 Cohort 1	Fellow 2 Cohort 1
	Fellow 2 Cohort 1	Fellow 5 Cohort 2	Fellow 9 Cohort 3	Fellow 1 Cohort 1
	Fellow 11 Cohort 3	Fellow 12 Cohort 3	Fellow 5 Cohort 2	Fellow 9 Cohort 3
(6-8) Middle School Teachers/Fellows	Fellow 6 Cohort 2	Fellow 7 Cohort 2	Fellow 3 Cohort 1	Fellow 10 Cohort 3
	Fellow 10 Cohort 3	Fellow 7 Cohort 2	Fellow 6 Cohort 1	Fellow 3 Cohort 1
	Fellow 3 Cohort 1	Fellow 10 Cohort 3	Fellow 7 Cohort 2	Fellow 6 Cohort 2
	Fellow 6 Cohort 2	Fellow 3 Cohort 1	Fellow 10 Cohort 3	Fellow 7 Cohort 2
	Fellow 7 Cohort 2	Fellow 6 Cohort 2	Fellow 3 Cohort 1	Fellow 10 Cohort 3
(9-12) High School Teachers/Fellows	Fellow 11 Cohort 3	Fellow 12 Cohort 3	Fellow 8 Cohort 2	Fellow 4 Cohort 1
	Fellow 4 Cohort 1	Fellow 11 Cohort 3	Fellow 12 Cohort 3	Fellow 8 Cohort 2
	Fellow 8 Cohort 2	Fellow 4 Cohort 1	Fellow 11 Cohort 3	Fellow 12 Cohort 3
	Fellow 12 Cohort 3	Fellow 8 Cohort 2	Fellow 4 Cohort 1	Fellow 11 Cohort 3
	Fellow 1 Cohort 1	Fellow 2 Cohort 1	Fellow 8 Cohort 2	Fellow 4 Cohort 1
Color key	District 1	District 2	District 3	District 4	District 5

6b **Induction ceremony highlights (How** did the induction ceremony go? Is there anything you would do differently? Please include up to 4 pictures with captions.)

Attendance at the induction ceremony (Please attach guest list with school district, government affiliation, etc.)

Media coverage of the induction ceremony (Please attach a copy of any media coverage.)

When do you plan on having your first meeting with the new Fellows? Describe your tentative agenda.

7 **End of the year (visitor) conference participation** (Did members of your team attend another site's end of the year conference? Please complete the tables below.)

Conference Location	Date of Conference

Participant's Last Name	Participant's First Name	Institution (district, university, etc.)	Role in the project

7a **Reflections on the end of the year conference** (Was the end of the year conference helpful for your team? In what ways could it be improved? Please provide specific feedback on each aspect of the conference.)

- **H-CCLS presentations**
- **Poster session**
- **Keynote speaker**
- **Other**

7b Do you think it is worthwhile for funds to be made available each year to attend another site's conference?

8 **Next year's calendar (Please provide an overview of your plans for next year. Include your calendar for next year.)**

a **Plans for next year**
b **Calendar (Please include a tentative schedule of meetings with Cohort 2, Cohort 3, and leadership meetings with DSCs and district meetings with Fellows.)**

Chapter 7

Innovation Phase of the Science Education Fellowship

Arthur Eisenkraft

I once saw a cartoon that depicted a father and son walking along the desert with a camel. The son asks the father, "Are we there yet?", to which the father responds, "We are nomads, we are never *there*." As we strive toward improved science education for all, I remind myself that we will never be *there*, but we must keep moving.

Through the Science Education Fellowship (SEF) program, we have started along the path of district transformation through teacher leadership. The SEF program's goal is to generate a critical mass of K-12 teachers within a district that can help create and support district initiatives. This requires the teacher-leaders to work alongside the science coordinator, the principals, and the district administration. We have seen many instances of this through the Growth Plan System (GPS) projects that the Fellows have pursued. We have also measured increases in their self-efficacy and seen them take on leadership roles as they speak up at school meetings, serve on committees, and provide professional development (PD) for their colleagues. As their role as teacher-leaders matures, they report more and more ways in which they continue to grow and to contribute to their schools and districts.

After the conclusion of the SEF program, we wanted to continue to support the journeys to district transformation. Too often, after funding ends, a program falls off a cliff where no further activity takes place and the district moves on to another program. The program certainly benefited the participants, the school, and the district, but the completed program may cease to exist. Having seen examples of this, we sought additional funding to continue the program and to wean the university and districts from the support while continuing to support the positive changes that took place. Figure 7.1 shows how adoption of a four-year program ceases after the funding ends and also shows how our plan is to continue the program with less and less external funding but

DOI: 10.4324/9781003490586-8

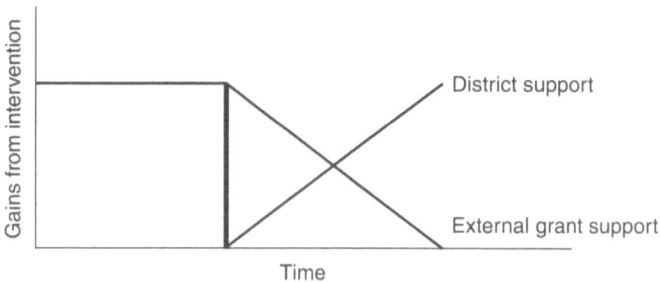

Figure 7.1 A sharp drop-off often occurs when external funding is cut suddenly. A weaning of external support and a stepping up of district support can help maintain program gains.

more and more district funding to maintain the impact over time with the sum of these two funding sources.

We began a new phase of the project, which we referred to as the SEF Innovation Phase and referred to our prior SEF work as the SEF foundation program. During the implementation of the SEF foundation program that has been described in detail in the preceding chapters, there was some variation across sites while there were also non-negotiables. For example, at one site where school districts had at least 100,000 students, we agreed that working with three districts made sense instead of the usual five districts. In another site that included rural districts with small student populations, we chose to work with more than five districts. In the SEF foundation program, the GPS year focus is split between personal and district goals. At one site, there was a big push to involve informal science organizations (e.g., science museum, botanic garden), and there, the Fellows in those districts split their GPS among personal, district, and informal initiatives. (An initial suggestion was to replace the "personal" with the "informal" but we reached the decision that three initiatives was a better approach than eliminating the "personal.") As an example of a non-negotiable, no site changed the structure of 20 Fellows for 125 hours per year to 100 Fellows for 25 hours per year.

As we embarked on the SEF Innovation Phase, we were challenged to come up with a four-year plan that would move us closer to school district transformation. The seven universities each came up with a plan that best suited the university and the school districts that they worked with. The plans were discussed, amended, and finalized. In this chapter, we will describe that process and those plans. First, however, we will summarize both the history of SEF and some of the successes of the SEF Foundation Phase.

Brief history of the Science Education Fellowship program

The SEF program has been generously funded, first by National Science Foundation (NSF) and then through Wipro. The Collaborative Coaching and Learning in Science (CCLS) work was initially part of the Boston Public Schools and the Boston Plan for Excellence. It began with a focus on math and literacy. In the 1990s, early content standards in ELA and math were promoted, and the Boston Superintendent, Thomas Payzant, established a school-based coaching program to support the transition to those standards. In School Year 2002, the Boston Plan worked with principals and teachers to pilot a new model for coaching, known as Collaborative Coaching and Learning (CCL), where teachers were provided with contractual time set aside for PD during the school day. A coach supported teachers as they studied, demonstrated, and mastered together an instructional strategy in ELA or math. They then translated that learning into practice in their own classrooms. As part of our NSF-supported Boston Science Partnership, we were able to adapt the strategies to science instruction to allow for teachers working across schools in the district – CCLS. Through a follow-up NSF grant, we were able to further articulate the SEF work with a focus on energy education.

After these Boston Public School-focused initiatives in collaboration with the University of Massachusetts (UMass) Boston, we were approached by Wipro, an international information technology company. Wipro Limited (NYSE: WIT, BSE: 507685, NSE: WIPRO) is a leading technology services and consulting company focused on building innovative solutions that address clients' most complex digital transformation needs. With more than 230,000 employees and business partners across 65 countries, they deliver on the promise of helping their customers, colleagues, and communities thrive in an ever-changing world. For additional information, visit www.wipro.com. As an employer with more than 15,000 US employees, Wipro was keenly interested in supporting a science education initiative. After discussing a range of possible grant programs, Wipro selected the Wipro Science Education Fellowship (SEF) as the program to champion, beginning in 2013. We redesigned the SEF model to be one that would be effective not only across schools, but districts, and across national sites as well. The financial support from Wipro would provide for stipends for each of the teachers of $10,000 over two years as compensation for their out-of-school work. (This sum was similar to the amount that NSF provides in their Noyce program and met the contractual agreements of teachers in Massachusetts and New Jersey.) Additional money in the grant went to facilitating the program and supporting Fellows' projects.

The Wipro SEF began as a school district transformation program to improve the quality of science education that K-12 students receive. What began at two university sites has expanded to seven university sites. During this time, the Wipro SEF program has positively impacted 35 school districts. These high-needs districts have been chosen to be geographically linked to Wipro sites. With this choice, the improvement in school districts takes place where Wipro employees work and live and where their children and neighbors' children attend school.

Through this public-private partnership, Wipro has made a lasting contribution to the communities, has become a corporate leader in promoting science education, and has provided a model for district improvements across the United States. They also created a university, school district, and corporate collaboration that should be emulated by other corporations.

Successes and the positive impact of the SEF

The Wipro SEF program creates a cadre of teacher-leaders who lead without leaving the classroom. These teacher-leaders (Wipro Fellows) make improvements in their science teaching and work with their colleagues to administer parallel changes in their classrooms. They also serve as an academic cabinet for the district science coordinators and principals in the district.

The positive impact of the Wipro SEF is spread across the seven universities involved in the program. Across these sites, the Wipro SEF has created a network of the

35 district science coordinators who share their expertise while challenging the status quo in their respective districts. The 420 Fellows have established working relationships among other Fellows at their site from other districts as well as with Fellows from other university sites.

I am often asked what has kept me a happy teacher and what has made my work meaningful throughout the years? Without a blink of an eye, I joyfully respond: The Wipro Science Education Fellowship! Since joining the program in 2015, I have grown as an educator and as a person. I have had many opportunities where I engaged in deep conversations that have led to meaningful work; all driven by curiosity, current trends and passions.

What makes this organization so powerful is it brings out the best in all. It reaches everyone where they are and lifts them up to see themselves as science leaders. The deep conversations, continuous professional development, and the encouragement of teamwork cultivates a close knit Wipro Science Community. The program is at different locations and consists of a wide range of people including K-12 teachers, professors, and administrators, which allows individuals to gain a wide perspective on science education. It also serves as a crucial reminder that you always have someone to lean on. Knowing that I have this support of people within my district and beyond has helped me gain a reputation of someone who enjoys helping others and someone who shares the joy of science. Additionally, it has served as a strong reminder: It takes a village. I now proudly serve as the District Science Coordinator of the Wipro Science Education Fellowship for the City School District of New Rochelle. It is very exciting and humbling to think about how far I have come in my science education journey, from doing science with my 20 students to now offering my expertise to other educators through organizing professional development opportunities, participating in district curriculum work, and leading projects that meet the needs of all. The number of Wipro Science Education Fellows has quadrupled in participants since I joined, too! In our latest project, the school psychologist is a team member and brings another perspective to our work and increases our projects' impact.

The Wipro Science Education Fellowship serves as a strong reminder of how an organization can run, where everyone feels valued and the work continues through deep reflection, ongoing conversations, and innovation. I look forward to continuing this exciting journey and cannot wait to discover the future hidden gems it will bring to all!

Aimee Ferguson, Elementary school, New Rochelle, NY

Referring to our logic model as a guide in Figure 7.2, we can see the progress we made through the SEF program.

The major focus of Wipro SEF is the guidance provided to teachers as they become teacher-leaders. The impact can be viewed at the school, district, and university level.

- At the school level:
  - The first year of the two-year program for Wipro Fellows is the focus on improving classroom instruction with a particular lens on equity. The Fellows chose research articles and applied that research to their teaching. The Wipro SEF program also supported cross-district collaboration, networking, and learning. Fellows developed a strong community within their cohort as well as with their own district teams across grade levels. Teachers had opportunities to work together

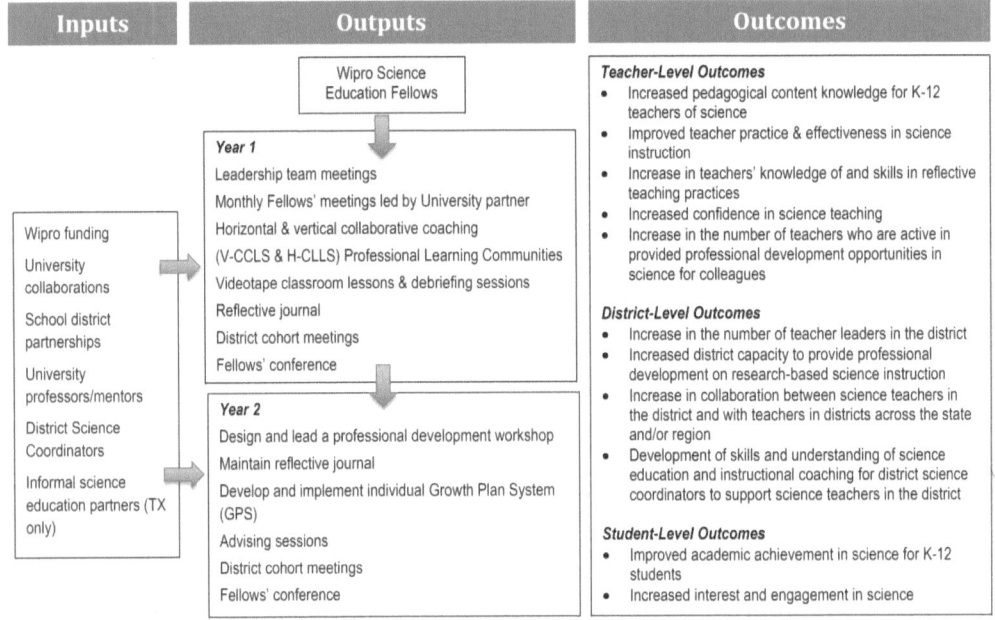

Inputs	Outputs	Outcomes

Wipro Science Education Fellows

Year 1
Leadership team meetings
Monthly Fellows' meetings led by University partner
Horizontal & vertical collaborative coaching
(V-CCLS & H-CLLS) Professional Learning Communities
Videotape classroom lessons & debriefing sessions
Reflective journal
District cohort meetings
Fellows' conference

Year 2
Design and lead a professional development workshop
Maintain reflective journal
Develop and implement individual Growth Plan System (GPS)
Advising sessions
District cohort meetings
Fellows' conference

Wipro funding
University collaborations
School district partnerships
University professors/mentors
District Science Coordinators
Informal science education partners (TX only)

Teacher-Level Outcomes
- Increased pedagogical content knowledge for K-12 teachers of science
- Improved teacher practice & effectiveness in science instruction
- Increase in teachers' knowledge of and skills in reflective teaching practices
- Increased confidence in science teaching
- Increase in the number of teachers who are active in provided professional development opportunities in science for colleagues

District-Level Outcomes
- Increase in the number of teacher leaders in the district
- Increased district capacity to provide professional development on research-based science instruction
- Increase in collaboration between science teachers in the district and with teachers in districts across the state and/or region
- Development of skills and understanding of science education and instructional coaching for district science coordinators to support science teachers in the district

Student-Level Outcomes
- Improved academic achievement in science for K-12 students
- Increased interest and engagement in science

Figure 7.2 The Wipro Science Education Fellowship Logic Model.

throughout the year as well as participate in professional conferences where they were able to share their work with each other.

○ In the second year of the Wipro SEF program, there is a stronger emphasis on developing teachers' leadership capacity. During whole group sessions, teachers are introduced to leadership concepts and explore what it means to practice leadership in their settings. Teachers are also asked to implement a year-long professional growth project aligned with their personal goals as well as their district goals. Teachers complete their time in the program with a stronger leadership identity and influence their school systems in ways that lead to stronger science classroom instruction.

- At the district level:

○ At the district level, there has been an increased understanding of how to support high-quality science teaching and learning. District Science Coordinators (DSCs) have developed strong relationships with Fellows within their districts as well as with the university staff. The DSCs and their Fellows have developed some common understandings and expectations with regard to high-quality teaching and learning, and district needs and initiatives. The Fellows engage in and provide support for the work of the district.

- Teachers in the Wipro SEF program have also become more involved at the district level with district science initiatives and work.
- DSCs have benefitted from their interactions with their site level DSCs and their learning together and supporting each other. DSCs are likely alone in their district and SEF affords them opportunities for role-alike engagement. District coordinators have benefited from networking among the seven Wipro sites during the leadership conferences. The 2021 online leadership meeting provided the unique opportunity for a large number of DSCs to meet without having to travel. They had time to talk leisurely, revisit colleagues over a month, and exchange ideas among different types of districts (urban, suburban, rural).
- One of the major impacts of the Wipro project is the creation of the teacher network, which has led to numerous collaborations among the teachers. This is particularly valuable to teachers in rural districts. Teachers in rural districts face unique challenges, primarily related to their isolation and lack of resources. A rural high school science department may have one or two teachers on staff, and they teach all science subjects. The Wipro SEF project provided these teachers with a network that created professional relationships that they leveraged in productive ways.

- At the university level:
  - Innovations to and outgrowths of the program have taken place at each university site including a center for STEM education at one site (New York), the expansion to informal education partners (Texas), inclusion of rural schools (Missouri), expansion beyond the Wipro SEF program to support large districts (California), work with an exceptionally large district (Florida), and various models for continuation grants (Massachusetts, New Jersey, New York).
  - The Wipro SEF program has been able to develop strong partnerships with local districts as well as with the seven participating universities across the country. The sites have learned much from the cross collaboration, including varying ways of approaching the work with teachers and districts in our own settings. We have become a strong community with a common goal of supporting high-quality teaching and learning in science.
  - In addition to the strong community, we have also benefited from the research opportunities that have been provided by the Wipro SEF program.
  - Wipro SEF has positioned participating universities to be prominent players in science education, supporting K-12 teacher development in large, local school districts.

- The positive impact of the Wipro SEF program is situated in the individuals benefitting from the program including:
  - 35 district science coordinators.
  - 480 teachers as Wipro Fellows.
  - At least 1,500 additional teachers who have participated in projects initiated by the Wipro Fellows.
  - More than 450,000 students during the eight years of the program.
  - Many, many other teachers who have attended workshops and/or read articles about the Wipro SEF program and projects.

The following are the key findings from the evaluation of the Wipro SEF since the external evaluations beginning in the 2015–2016 academic year:

- **High program satisfaction** – A large majority of Wipro SEF Fellows have consistently reported being satisfied or very satisfied with their Wipro SEF experience (ranges from 73% to 98% depending on year and subgroup).
- **Improved science teaching and leadership skills** – There is consistent evidence that Fellows involved in the Wipro SEF show improved instructional practices (particularly science teaching practices), science content knowledge, and leadership skills. Specific skills commonly showing significant growth in two or more years since 2016 include:
  - Confidence in science-related content and practices.
  - Leadership self-concept and leadership communication behaviors.
  - Reflective practices.
  - Use of research to guide their professional practice.
  - Confidence with the Next Generation Science Standards (NGSS).
- **Self-concept as teacher-leaders** – At the culmination of the Fellowship, most of the Fellows see themselves as teacher-leaders in their schools (as high as 86%) and some in their districts (as high as 26%).
- **Successful program elements** – Fellows have continually recognized the importance of:
  - The collaborative network of science educators created by Wipro SEF.
  - The specific value in both horizontal and vertical collaborative work.
  - The value of the GPS experience.
- **Program hallmarks for district administrators** – DSCs repeatedly noted several key impacts of the Wipro SEF program for their Fellows:
  - A growth in teacher leadership and confidence.
  - Improvement in science teaching skills.
  - Continued growth of a collaborative network of science educators.
- **Positive impact on districts** – DSCs and district administrators in California, Florida, Missouri, and Texas and former Fellows in the Northeast commonly note a number of impacts on districts involved with Wipro SEF including:
  - The value of a supportive and collaborative network of science teachers.
  - A growing awareness of the importance of science at the district level.
  - The creation of and improvement of teacher-leaders.
  - An improvement of science teaching and access to science for students.
  - A highlighting of the importance of science at the district level.
  - An energizing of teachers, including veterans.
  - The value of opportunities for vertical collaboration.

Over the years, Fellows have often shared heartfelt comments that convey the importance of the Wipro SEF in their professional lives. Below are a few noteworthy comments shared in program evaluations (anonymously) over the years:

- *"Wipro was without a doubt the best professional development experience I have ever participated in." (Fellow, 2021)*

- *"I feel like we received such a great balance of challenge and support. The requirements of the program were definitely challenging, in a good way that pushed me to reflect and grow as an educator. That being said, the program really supported us in moving at our own pace, from our own truths, and through our unique hardships during this pandemic year." (Fellow, 2021)*
- *"Wipro was undeniably the single greatest professional development I have ever participated in. It has improved almost every aspect of my teaching and has fundamentally changed how I think about teaching." (Fellow, 2020)*
- *"The way in which I benefited the most was that the program energized me. I was excited to try what we learned and discussed in my classroom with students. To have a group of teaching professionals all sharing a commitment to education and a commitment to improving their practice motivated me. This carried over into the classroom. This program helped me to incorporate NGSS into my classroom and put it into practice at an entirely different level." (Fellow, 2020)*
- *"I feel that collaboration with other educators both vertically and horizontally helped me become a better science teacher." (Fellow 2019)*
- *"I was able to network with like-minded individuals that share my passion for science education. This was, without a doubt, the most valuable part of fellowship. I'm thankful for the opportunity to work with others, brainstorm, learn, and grow." (Fellow, 2019)*
- *"My eyes have been opened to the research available to me to. I have become more reflective. I have learned so much about teaching science and have found many ways to integrate what I have learned to other subjects." (Fellow, 2018)*
- *"My participation this year in Wipro SEF was a great experience because I was able to research a strategy that I was interested in. The results of my research demonstrated to me how important it is to create more meaningful experiences for the students and to create an environment that involves many opportunities for collaboration." (Fellow 2018)*
- *"I was able to change my perspective in science teaching, see where the gaps are in my own district, and hopefully make some really important changes in regards to science and how we are providing instruction to our children." (Fellow, 2018)*
- *"I learned more about myself through Wipro SEF. I now have the tools to conduct myself more professionally when collaborating with others. Analyzing articles and trade journals, as well as participating in science discourse, allowed me to re-examine how I plan and facilitate lessons. Now, I have a more critical eye when deciding to implement hands-on activities in my classroom." (Fellow, 2017)*
- *"I was able to invest my time in a program that I was interested and passionate about. It allowed me to be more of a leader in the building working with other teachers and students to get the program running. Being a part of Wipro SEF also has given me opportunities which I would never have had otherwise." (Fellow, 2017)*

Parallel comments from DSCs, principals and Institution of Higher Education (IHE) professors and staff complement those of the Fellows:

DSCs

- *"I really enjoy the camaraderie of the other participants and the learning that we do as a group. The facilitation and support is just incredible." (California DSC, Year-end Survey 2024)*

- *"I gained knowledge related to the strategies implemented through each of the fellow projects. The discussions, articles read, and research shared contributed to my professional growth." (Florida DSC, Year-end Survey 2024)*
- *"The National Wipro district coordinators conference was amazing. I learned specifically how to leverage the work teachers have already done to collaborate around how to bring the work district wide. Wipro has helped me build the Science Teachers Leadership Team, which are working on action plans to connect the work they have done with district initiatives." (California DSC, Year-end Survey 2022)*

Principals

- *"I really enjoyed and appreciated being a part of the School Leader's Cohort. We had conversations that were so different than what happens every day, so much richer. It was a great opportunity to meet and work with other administrators – a unique program." (CA Elementary Principal, Interview 2024)*
- *"The SEF program has been a great networking opportunity as a principal. I feel really connected to UNT Dallas. It has given us opportunities to present. I wish there were other programs like this that take care of their people and really make you feel known and like you're making a difference." (TX Middle School Principal, Interview 2024)*

IHE

- *"I am so grateful for the Wipro SEF family. I wouldn't trade it for anything. And I'm so grateful for what I've learned." (CA IHE, Interview 2024)*
- *"Our doctoral team has just flown with this Wipro SEF research work. I love to see more Fellows publishing and presenting at conferences. It energizes our whole team!" (NJ IHE, Interview 2024)*
- *"I appreciate being a part of a community of practice with the other IHE faculty. The in-person meetings are very valuable." (NY IHE, Interview 2024)*

Wipro Fellows as well as DSCs and university professors associated with aspects of the program have conducted conference presentations and published articles in the media, teacher practitioner journals, and research journals. Over the past eight years, we have shared our Wipro SEF work in a variety of ways, from board meetings to scholarly presentations at 20 national and international conferences, as well as through scholarly publications in eight peer-reviewed journals and three chapters in books. Many of the presentations included Fellows and DSCs, where together we reported research on the Wipro SEF model and have shared the innovative PD model with other teacher educators.

Success breeds success. Some of the Wipro Fellows have been able to secure additional external funding to further their innovative work in their schools. For example,

- A Fellow received a $1,200 grant from the National Future Farmers of America to cover the costs of a school bus, organic soil, and vegetable seedlings that his students will install at the University of South Florida Botanical Gardens.

- A Fellow was chosen as a finalist to win $10,000 dollars for a greenhouse and chicken coop at their elementary school.
- A Fellow received a $55,000 grant from Loews to create Maker Spaces in her elementary school.
- A pair of Fellows received a $500 grant from a local company after writing about their Wipro experiences and used the funding to select and purchase a new science curriculum for K-5th grade.
- A Fellow received the 2021–2022 STEM grant from her district, which provided $2,000 to equip all 4th grade classes with STEAM carts and activities to implement STEM within their classrooms.

Detailed documentation of the successes and growth of Wipro SEF are in the quarterly and annual reports over the past ten years (available on our website. https://wiprostem-program.com/).

The Innovation Phase

To continue to move forward, each IHE was charged with creating a new set of initiatives that could build on the SEF foundation program and help move us toward our goal of district transformation through teacher leadership. As seeds for thought, we arrived at some broad-stroke approaches from which each site could develop a proposal.

Proposed projects for this round of continuation funding

A   **Present sites** – Build on the expertise of the present universities and school districts and cadres of Fellows.

   a   Continue the program with new cohorts with the present model. These could be with:
      i   The present large districts to reach a critical mass.
      ii   New districts in the surrounding areas.
      iii   Expansion to rural districts within present sites' states.

   b   New models with the present districts.
      i   Wipro SEF for DSCs and other administrators (e.g., principals).
      ii   Wipro SEF for math and/or interdisciplinary teams (SEF = STEM education fellowships).

   c   Supporting post Fellows work within districts.
      i   Expansion of successful GPS projects from Year 2 of Fellowships including expansion of work to include other teachers.
      ii   Setting up CCLS within their districts with other teachers.
      iii   Targeted mini-grants to work with other districts.

   d   Other routes for new teachers to become Wipro Fellows.

   e   Expansion to include other non-profits (e.g. Talk STEM in Dallas focusing on women/minorities).

B Cross-site initiatives

    a Working together through conferences and sharing.

        i DSC conferences.

        ii Principal conferences.

        iii Teacher Leadership support (articles, presentations, presentations).

    b Evaluation and Research.

        i Continue evaluation across sites.

        ii Research efforts building on Teacher Leadership including how our Fellows develop and implement teacher leadership; how the Fellows define teacher leadership through their choice of GPS projects, and their interactions with teachers and their districts; and how the distributed leadership model within districts has changed.

Each site submitted a proposal and, through follow-up discussions including budget considerations, arrived at a plan for the four years following the SEF Foundation program. The programs are summarized next to demonstrate the breadth of chosen paths. Please note that because of the rollout of the SEF Foundation program in different years, some of the sites have been in the Innovation Phase for more years than others.

California

The California site at Stanford University chose to add two new cohorts of Fellows as well as initiate a program for school leaders.

By recruiting two more cohorts of science teachers from the existing five partner districts to participate in the two-year Wipro SEF Program, the strong relationships that have been established between the IHE Center CSET (The Center of Support for Education and Teaching) will grow and the critical mass of Fellows in each district will increase. The goals and pillars of the program will remain fundamentally the same as the current (Foundation) program. The Bay Area is densely populated and working with the same districts offers the opportunity to more strategically strengthen science leadership in this area by going deeper in each of the districts. We will be investigating the value-added from enlarging this pool of teacher-leaders in each district. In addition, the California site is investing in district teams from the five partner districts consisting of DSCs and Wipro Fellows. We are working more closely with each district team to develop their collective capacity to advance effective science teaching and learning in their districts that highlights the NGSS and reduces the persistent inequities that pervade science education. This not only requires working with DSCs and Fellows, but also requires more direct involvement from principals who have remained mostly in the periphery of the Wipro SEF program. Thus, the California site is developing a program specifically for school leaders with the aim of creating strong district teams that can make transformational changes at the site and district level. The Wipro School Leaders Program launched with a four-day summer institute. Eleven school leaders met during this time to consider how they are leading their schools and building learning communities. During the school year, the program will shift toward science instructional leadership.

Finally, the California site is supporting the conceptualization and implementation of continuous professional learning experiences for DSCs from all Wipro sites. For this work, the California team is working in collaboration with UMass Boston to implement leadership conferences and virtual sessions over the course of three years.

Florida

Selected Fellows will submit an enhancement project that will be used to increase the impact of their work across the district. They will be forming teams, which must include an official leader of some type and another teacher previously not affiliated with the project. They will submit for funding for either a one- or two-year project. As part of the project, they will be required to present both at a Wipro-sponsored event as well as within or beyond the community, and for the two-year projects, they will need a national presentation/publication of some type. They will learn how to conduct action research in addition to how to disseminate their findings. The four main categories for funded projects are as follows:

- Equity and social justice in STEM.
- Technology enhancements.
- Advanced curriculum development.
- Establishment of a new Vertical Collaborative Coaching and Learning Science (V-CCLS) group.

Examples of the projects underway are:

Title: BSCS 5E Instructional Model at Jule Sumner High School

This two-year project aims to deepen participants' understanding of the BSCS 5E Instructional Model to support planning for instruction and assessment aligned with the Next Generation Sunshine State Standards (NGSSS) and A Framework for K-12 Science Education. The project's goal is for participants to learn how to develop phenomena-based 5E instructional sequences to support coherent storylines and conceptual flow aligned with the NGSSS and A Framework for K-12 Science Education.

Title: Working Across Grade Levels to Improve Grades 3–5 Science Teaching

This two-year project brings together grade 3–5 teachers in a CCLS to improve the teaching of science at their school. The team for this project is meeting bi-weekly to look at their standards and curriculum across grades 3–5. Since the state of Florida doesn't have a clear vertical progression of the science standards, we will work together to discuss the content knowledge the students should have received in previous grade levels, what they need to know in their current grade level, and what they will learn in the next grade. They are looking closely at the grade level standards by Big Idea (Florida Standard) and identify vocabulary that students need to know for each grade level. They then identify where the foundation begins for concepts we feel students have a difficult time grasping, and strategies and resources we can use to help fill in gaps or to help to progress their knowledge on a particular concept.

Title: Gifted but 'Off Track': Serving the Gifted Students of a Title 1 High School Team

In this two-year project, a Phase I Fellow, a classroom teacher, and the assistant principal, are establishing an after-school club to support gifted students who have been designated either "at-risk" or "off track" according to Early Warning Intervention data.

Massachusetts

The UMass Boston innovation plan includes working with the original five districts as well as beginning Wipro SEF activities with three new districts.

Each of the original five districts had a meeting with the Fellows, the DSC and the IHE to discuss district initiatives. The purpose of these meetings is to help define "district transformation" for each district. This requires identifying the gap between the present district situation and the future vision of the district. This leads to a recognition that there are specific changes that the district may want to implement in science. In turn, we identify strategies that are within the capabilities of the Fellows (i.e., teacher-leaders) and the coordinator to implement over the next few years. The districts will propose a plan and the support (instructional and financial) that are needed. Once the Fellows, coordinator, and IHE agree on the plan, the coordinator and Fellows introduce the plan to the administrators (principals and assistant superintendents) for their input and approval. Once the administration gives the go ahead, the district formally applies for a Wipro grant. When the grant is approved, outreach in the form of newspaper articles and school board presentations will take place to celebrate the success of the district in having a plan and grant approved.

UMass Boston will also try to generate interest in the foundation Wipro SEF program of four years in three high-needs districts in the Boston area. Unlike the original sites, these new sites and Fellows will not be receiving the generous stipends of the past and will have to come up with other ways to incentivize participation in the program.

Missouri

The Innovation Phase project at the University of Missouri will expand the teacher network, provide new opportunities for leadership, and focus on collaboration between science and math teachers in middle and high school. Middle and high school teachers from each district will enroll in the Wipro SEF project as teams of two to four teachers, with each team having a math and a science teacher from the grade band. Three cohorts of 15 teachers each will be recruited, with each teacher participating for two years. The project will address the challenges of teaching science and math in a harmonious manner at the middle and high school grade levels. The collaboration between math and science teachers is essential to the implementation of a successful science curriculum. Teachers will collaborate in V-CCLS and Horizontal Collaborative Coaching and Learning Science (H-CCLS) teams and devise an implementation plan for restructuring their science/math instruction at the end of Year 1 to be more compatible. Working together with their DSC and school administrators, they will define how wide the restructuring will be, how it integrates with the district goals, and how they will evaluate it. They will expand and implement the changes in Year 2 and report back on their successes and challenges. They will share their plans within and beyond their cohorts. They will be encouraged to disseminate their learnings through professional development, conferences and/or websites.

In Year 2 of each cohort, elementary teachers will be recruited from Year 1 Fellows' districts as associate Fellows. Elementary teachers typically teach both math and science. The purpose of having them work with middle and high school teachers is to have them learn content and methods as well as work on vertical collaboration across the K-12 spectrum so that they can integrate science into their math classes and vice versa.

New York

The Innovation Phase of Wipro SEF generated teacher-driven project ideas that could be supported. Many Fellows continue their Wipro work even beyond the scope of the foundational program. Some Fellows applied for mini-grants multiple times, some continued important interdisciplinary work, and others continued to conduct PD for teachers in their districts. It is especially exciting to see Fellows participating in other PD opportunities provided by the IHE Center (MCSE (Mercy Center for STEM Education)). Since Phase I, Fellows have continued to be part of the MCSE family by participating as Saturday STEM Academy instructors, Westchester STEM Ambassadors, and presenters at the annual K-12 STEM Teacher Conference.

To continue the work of Wipro Fellows in this new iteration, MCSE aims to support teachers and Fellows in a new round of GPS projects. These projects will be collaborative within districts and receive buy-in from administrators as associate group members, as well as in-district support from DSCs. Over the course of four years, MCSE aims to establish a norm of collaborative action toward district change in our existing Wipro districts. With the focus on sustainable change, the plan is to equip participating teachers with the tools and practices necessary to carry on transformative efforts even when Wipro funding is gone. Furthermore, MCSE will insist on administrator buy-in with the support of DSCs to set the foundation for transformation and to hold district administrators accountable for this level of change. This project will require steps at each level of the district to ensure sustainability.

Examples of some present projects are:

District	Project Title/Description
Port Chester	Edison's Kindness Garden – Provided green spaces for scientific observation, inquiry and experiments. Engaged K-5 students in every stage of the garden project: design, planting, caring and harvesting. Planted a variety of sensory-rich vegetables, herbs, and plants in thematic raised garden beds. Increased parental engagement and provided an opportunity for the entire learning community to experience the joys of gardening activities.
New Rochelle	Vertical Integration of STEAM in Elementary School – The purpose of this project was to build a foundation of engineering vertically and horizontally using an interdisciplinary approach. The goals were for the students to be able to understand the engineering method: designing, researching, hypothesizing, testing, and drawing conclusions.
Port Chester	Eggceptional Bridges: 4th and 5th Grade Engineering Investigations – Through a hands-on approach, students engaged in experiments, creating hypotheses and testing their theories with various materials. Fourth- and fifth-grade students worked with teachers over six months and explored material science and basic engineering. These lessons gave students a greater understanding of the materials and engineering needed to build a small bridge for a shared school garden.
White Plains	STEM Hub: Authentic Experiences in Science and Engineering for Young Learners – A group of educators created authentic STEM investigations that are culturally and historically responsive using pedagogy of Dr. Gholdy Muhammad, Dr. Eugenia Etkina's ISLE model, and the 5E instructional model for inquiry teaching.
New Rochelle	New York State has new Computer Science and Digital Fluency standards. As a result, the Jefferson Wipro Team thought about ways to bring all stakeholders up to speed by considering these standards while infusing Social-Emotional Learning. This group hosted tinkering events, professional development on standards, reading materials for adults and children, assemblies, lunch clubs, and parent workshops.
New Rochelle	This group expanded an existing program for K-1 graders (FLORES) to include grades 2 and 3. This series of workshops brought together families in the Columbus school community with the objective of empowering parents to become science facilitators and to excite students about more advanced coding, engineering, and vermiculture.

New Jersey

Based on research findings, in this Innovation Phase, Montclair State University is leveraging partnerships with their current districts to expand the program to new districts and teachers. In line with a major goal from the Foundation SEF and their first Innovation Phase, they are supporting experienced Fellows to continue to flourish as teacher-leaders and to become leaders for the program. The specific plan includes:

• Recruiting two cohorts of Wipro SEF Fellows over four years. Each cohort will participate for two years and will include 15 Wipro SEF alumni and 15 new Wipro SEF Fellows.

- Supporting the alumni Fellows to undertake project that

  - Is GPS-like in nature and supports a district initiative and helps them grow as teacher-leaders. They would do this in collaboration with a new Fellow who they recruit.
  - Helps expand Wipro efforts by recruiting new Fellows. For example,

    - A teacher from a different district/school.
    - A teacher from another subject area (e.g., math or special education).
    - An administrator or non-faculty colleague (e.g., librarian or curriculum coordinator).

This model builds on the foundation set in earlier phases and expands and disseminates the model in a way that is both manageable and sustainable. The overarching goal is to support our experienced Wipro SEFs as they continue to deepen their leadership capabilities and branch out to new teachers and districts.

We anticipate that this model of expansion will lead to the following outcomes:

- Stronger university/district partnerships that includes an expanded list of districts.
- Participation of up to 60 new teachers including those in other content areas.
- Support of district initiatives.
- PD experiences in the areas of diversity/equity/inclusion; interdisciplinary curriculum and pedagogy (particularly focused on science and math); teacher leadership.
- Continued engagement of Fellows with larger professional community at all levels (school, state, national, global).
- Continued dissemination of Wipro SEF model and resulting research.

The current projects being undertaken by the Fellows range from V-CCLS/H-CCLS type activities to partnerships with local community programs. A snapshot into the Fellows' work includes:

- Expanding Data Literacy and Increasing Collaboration Among Math and Science Teachers
- Gaining The Buy In – Teacher Leadership
- Assessing Science PD needs
- STEMtastic Student Engagement Liaisons Seeking Solutions
- Follow the Light.
- School 17 STEAM Club
- Removing the Barrier of Language from Science Instruction
- Facilitating Math Stations in the Elementary Classroom
- Wipro Number Strings
- Getting Involved through STEAM
- Garden Expansion Project

Texas

In the first Innovation Phase at the University of North Texas, Dallas, they were able to support Fellows in a wide range of projects and disseminate their work at conferences. Highlights included:

- The WalkSTEM project for the DSCs and a partnering teacher from all 5 ISDs.
- The Southwest collaborative Online PD conference in June.
- Seven collaborative/mini-grants were successfully completed.
- Each of the eight projects sent a proposal to be presented at the Conference for the Advancement of Science Teachers (CAST) 2022 and five proposals were accepted for presentation at the conference. CAST is an annual regional conference hosted by the Science Teachers Association of Texas. It is a requirement of each project that all participants submit a proposal to present at CAST or a conference of their choice.
- **Project Titles**

  All Hands-on Deck: The Impact of Hands-on Activities on Science Instruction
  Science Text Comprehension Through Note Taking
  STEM & Informal Writing Tasks Build Writing Skills
  STEM Collaboration Across Classrooms and Beyond
  Using WalkSTEM to promote student inquiry in the real world
  Science! It's Elementary!
  Investigating Climate's Impact on the Environment

In their next Innovation Phase, they will have three components targeting different groups of science educators and students.

1 The School Projects, focused on one school per collaborating district per year.
2 The DSC collaborative project where each DSC and an Irving Independent School District (ISD) teacher partner work together with other DSCs and their teacher partners.
3 The Collaborative/Mini-grants led by Wipro Fellows, collaborating with Wipro and non-Wipro teachers within or between schools and district.

These are examples of some of the projects that are underway:

I	School projects				
	District	*Title and focus of project*	*Grade level*	*New Fellows*	*DSC/ Alums involved*
a	Lancaster ISD	5th Grade Science Teacher PLC	5	1	1 DSC participant 2 Alums
b	Cedar Hill ISD	Effects of Collins Writing in 8th grade Science	8	4	1 DSC participant

(Continued)

A Roadmap for Transformative Science Teacher Leadership

I	School projects				
	District	*Title and focus of project*	*Grade level*	*New Fellows*	*DSC/ Alums involved*
c	Irving ISD	STEMing to Staar	5	0	2 Alums 1 DSC advisor
d	Grand Prairie ISD	Which Properties Matter?	2–3	3	1 DSC advisor
e	Grand Prairie ISD	GFAA ST*Arts* Club! students	3–5	3	1DSC advisor
f	Desoto ISD	Preparing students for Staar 2.0	6–8	1	1 DSC Advisor
g	Grand Prairie ISD	STEMtastic Morning	6–8	4	
h	Grand Prairie ISD	Edible gardening	K–5	2	1 Alum
i	Grand Prairie ISD	NSEC Enrichment for middle school	8	2	2 Alums advisors
II	Collaborative projects				
a	Irving ISD Lancaster ISD	Exploring Science concepts using social studies in a cross curricular research study	9	1	1 alum
b	Advantage Academy Lancaster ISD	All Hands on Deck: Importance of Hands-on activities for Science Instruction	5	1	2 Alums
III	Individual projects				
a	Lancaster ISD	Science Staar Bootcamp 2.0	5		1 alum
b	Denton ISD	Classroom Educational Website for Science content	5		1 alum
c	Duncanville ISD	I CER You	Honors biology		1 alum

Summary

The varied pathways described in this chapter speak to the commitment of the universities and school districts to move forward toward district transformation. We can see examples of teachers in a school collaborating as well as schools in a district forming partnerships. We can see movement toward distributed leadership with the involvement of teachers, DSCs, and principals in other projects.

It is worthwhile to note that the administration of the SEF program has had positive impacts on the participating universities as well. Although we set out to move school districts forward, it is gratifying to see how the IHEs have moved forward as well.

Mercy College (now Mercy University) became one of the SEF sites in early 2014. Our leadership of the NY site, along with a concurrent National Science Foundation award, positioned us in the NYC metro region as a prominent player in science education, supporting K-12 teacher development in large, local school districts. Our expertise grew in this area as the SEF continued, as did our local reputation. We were able to leverage our contacts and esteem, as well as our external funding streams, to establish the Mercy College Center for STEM Education (MCCSE) in 2016. This MCCSE is the umbrella under which we serve the community through programs for children, families, and teachers. Our reach and impact have been greatly influenced by our participation in the Wipro SEF.

Meghan Marreiro and Amanda Gunning

The larger SEF community

The original SEF program model had a funder (i.e., corporate sponsor, foundation), a lead university, and five local public school districts selected to be part of the program. Fellowships were awarded exclusively to K-12 science teachers from these five partner districts. The lead university received additional funding and more universities were added that each partnered with their own local public school districts. The SEF program was implemented through each partner university, with the lead university coordinating the collective efforts of the university partnerships and SEF program implementation. The lead university provides quarterly reports to the funder and is the overall coordinator for sharing of information, meetings, and high-level interactions between all sites involved. See Figure 7.3 for a depiction of this cross-university partnership for the purpose of implementing the SEF program.

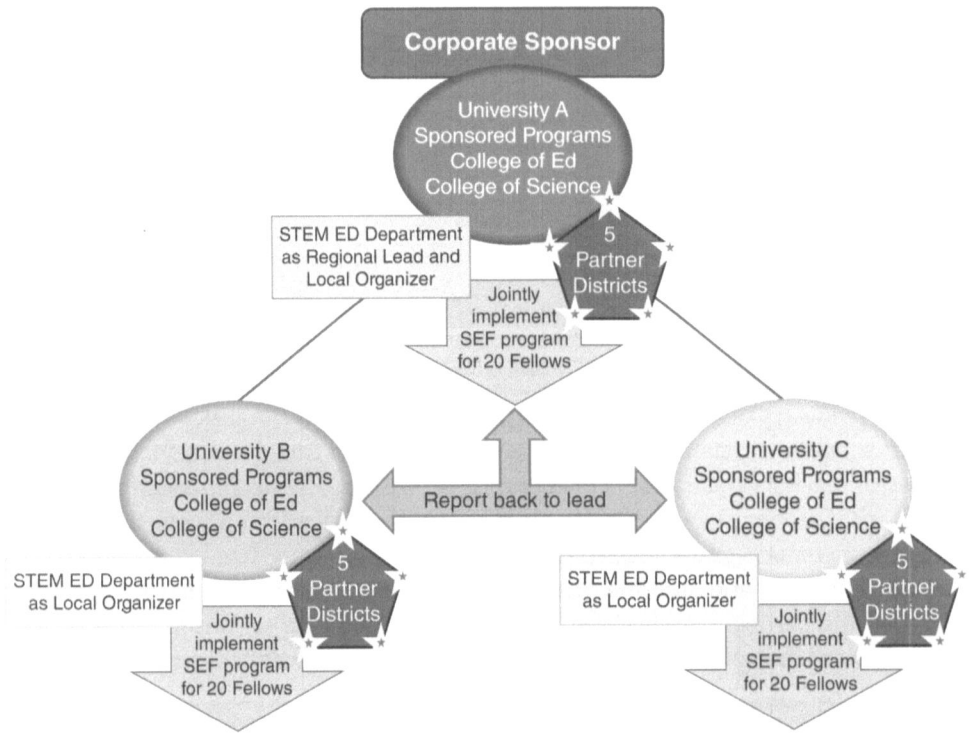

Figure 7.3 A depiction of a cross-university partnership to implement the SEF.

As other universities and districts come on board, it is essential that the core components and integrity of the SEF program are maintained. The program has certain features that we have found yield success. It also ensures a common language so that Fellows across sites recognize the program at other sites. At cross-site conferences, all participants can actively share in presentations about the V-CCLS and the H-CCLS while using the tuning protocol, as discussed in Chapter 3. The strong connection between universities helped the leaders of the SEF program learn from each other to ensure that the program experience for Fellows was high quality.

However, variations are to be expected as each university has its own context and needs. For instance, a university may partner with fewer (or more) than five districts, depending on district size. The funding and incentives for teachers to participate in the program may come from the districts (using monies set aside for PD or other non-monetary incentives such as release time) or an external grant rather than a corporate sponsor. Modifications of some features will not diminish the overall effectiveness of the efforts and may even enhance the collective efforts of the universities.

The lead university of the SEF program is the university at which the Principal Investigator/Director of the regional program resides. The lead university takes on the following broad responsibilities:

- To recruit and subcontract with other universities to run concurrent SEF programs.
- To manage the overall program and ensure the fidelity of the implementation of the SEF program at the partner universities.
- To identify an outside program evaluator
- To establish the roles and responsibilities of an advisory board.

The value of having several university sites running concurrent programs is that it allows collaboration among sites that serve school districts with varied demographics, for example, rural vs. urban, large vs. small, high vs. low socio-economic status (SES), and varied amounts of ethnic and racial diversity. Having a single evaluator allows for a comparison across these variations. Furthermore, the garnering of funding rests with the lead university, but is supported by partner universities, and can be a factor that is attractive to funders. Dissemination of the project across the country or state is built into the program. Collaboration occurs among faculty across the sites, among Fellows from different districts, and different sites as well,

As with any program implemented across several sites, it's vital to keep track of the variety of meetings and events that occur with each cohort, each year of the program. The SEF project has used two types of meetings across the sites – monthly video meetings and bi-annual face-to-face meetings. The purpose of these meetings is to help each site stay on track and synchronized with the activities of the other sites, while also discussing success and challenges and receiving support and advice from both the lead university and partners. These meetings are attended by the director and site leaders, and often by other members of partners' leadership teams.

During the monthly meeting, typical discussions include the previous month's activities, information about questions from the funders, and general progress. Activities coming up in future months, such as the annual conference, are also discussed. New ideas are often brought up and their logistics are discussed; for example, when the

COVID-19 pandemic moved all meetings, including the conference, online, the logistics of conducting the conference online, as well as ideas about having individual online conferences vs. one large conference across sites, we discussed in great detail (and yes, we tried both in successive years). Several ideas for supporting district science coordinators and providing PD for them came out of these meetings. Thus, the meetings provided not only a check-in but served as a thinking space among partner university leaders, who have, in the process, become close colleagues.

During the bi-annual meetings, typically held in early fall and in early spring, one of two members of the leadership teams from all partner universities and the director of the lead university meet face-to-face at a host university. The communication between the different universities' SEF programs ensures fidelity to the program as it was intended and allows for a sharing of new ideas that can be tried across the site with an eye toward improving the program. For example, work on this book was finalized during a one-day workshop during one such meeting.

Set a research agenda

The SEF program is an ideal program upon which to conduct research to share with the academic and practitioner communities. The program seeks to enact meaningful, positive change in school districts through teacher leadership. There are thousands of school districts across the United States and even more internationally that could benefit from learning about what happens when the SEF program is enacted with fidelity. Those currently involved with the program have presented dozens of papers and posters at conferences and published more than 20 research articles in peer-reviewed journals and invited papers and presentations at conferences about the program. The beginning of a program is an ideal time to set a research agenda. A data collection plan can be developed and begin with the program rather than be retrospectively applied. It is certainly understandable to not enact a broad research agenda at the outset. Still, a focused research question to be explored from the beginning could lead to sharing of knowledge that will benefit many beyond the local deployment of the program. Key factors to consider during this research study are how will it benefit the participants and how the data collection can occur with minimum disruption or burden to the participants. There is a potential for "survey fatigue" in a program of this size and scope. A well-planned research agenda begun early in the program will help with fatigue among participants. Some teacher-Fellows like the idea of contributing to peer-reviewed research and may help with literature review and manuscript creation.

Creating and maintaining an inter-site community

While the SEF project is conceived around creating a community of teacher-Fellows and university faculty at a given partner site, as the SEF project developed partners recognized the need and the opportunity to build a broad multi-site SEF community among partner university leadership teams, among the district science coordinators, among the teachers, and among all these groups.

The inter-site community among the partner university leaders was possibly the earliest community created. The monthly online leadership team meetings and biannual face-to-face meetings produced an obvious forum for exchange of ideas, open airing of successes and challenges – and collegial feedback that proved helpful. These meetings were also the forum for other community building ideas, such as workshops for DSCs and content workshops for the broad inter-site community, including alumni Fellows and faculty. During the COVID years, which hit during the second year of the SEF program for four new partner districts, discussions of how to handle meetings, conferences, GPS advising, and morale among all members was openly and repeatedly discussed.

The annual conferences held at each site was a rich opportunity for cross-site visits by SEF Fellows, faculty, and DSCs. Fellows from visiting sites made presentations of their work, both H-CCLS and GPS, along with Fellows from host sites. In the first year of funding, DSCs and faculty saw where their site needed to be a year hence. In later years, interaction among Fellows from different sites gave them a sense of the national reach of the program. There have been instances of long-term friendships created among Fellows across sites, and instances of GPS projects based on keynote speakers' work across sites. These inter-site visits are a permanent fixture of the program.

Workshops based on specific topics, such as climate change, gardening, and English as a second language have been gaining interest, primarily because Fellows from different sites have chosen similar topics. The team has been working on conducting these cross-site workshops for Fellows, DSCs, and faculty. A small group of Fellows was also invited to conduct a workshop at a series hosted by the STEM Education Leadership Network and funded by the NSF. While these are examples, it provides a space where Fellows from several sites can interact with a national audience. Fellows have also presented their SEF work at national conferences. On a more informal level, lunches have been organized at these conferences to bring inter-site participants together.

The leadership conferences for DSCs morphed from face-to-face to virtual during the COVID-19 pandemic. This turned out to be a blessing in disguise, since it offered an opportunity to broaden the scope of the conferences as well as providing the convenience of not having to travel. The process used in 2021 is described next.

The UMass Wipro faculty team, with input from faculty from all sites, chose a variety of session topics. Five, two-hour sessions were held over a three-week period in February–March 2021. The leadership conference attracted 25 DCs and 12 Wipro faculty from various sites. Each session focused on one topic:

1   Sharing District Science Coordinators' Expertise, Knowledge, and Experience
2   Building Bridges & Community with Wipro Fellows
3   Lessons from COVID
4   Planning for the End-of-Year Conference and Beyond
5   Building Leadership Capacity

Session leaders were assigned – typically one faculty member and two volunteer DSCs. Session leaders met online prior to the conference sessions and created a detailed discussion schedule with three to four discussion items and a method of collecting responses. Each session had an anchor document with links to presentation slides and response documents. Participants brainstormed discussion items during breakout sessions of four to six attendees, and recorded their responses on a variety of platforms, including Jamboard, Google Docs, and Padlet. An outside team sat in on these discussions and created a summary graphic of each session (see Figure 7.4). DSCs who attended and led sessions were provided a stipend. Similar conferences were held in 2022 and 2023.

Figure 7.4 Graphic notes image.

Visual summaries by Pushpin Visuals.

Conclusion

We have covered a significant amount of material. There is a lot to take in. We have shared with you much of the heavy lifting for structuring the SEF program. Executing the program is up to you. There is no one way to roll out this program. You may want to start small with one school district and focus on vertical curriculum articulation using the V-CCLS process described in Chapter 3. Perhaps you are a new department chair or lead teacher and notice that your department/grade group approaches the science curriculum in vastly different ways. In this case, perhaps the horizontal application of science and engineering practices and the H-CCLS process is for you. You may find that your district has already invested significantly in both vertical and horizontal alignment, but you need to nurture the next generation of leaders. GPS projects like those described in Chapter 4 may be for you. You may also find that you are ready to dig in and build a program with significant reach like the original SEF. Regardless of where you find yourself, this book and the resources within it will help you get started. Combine that with your own initiative, skills, and the curiosity that led you to pick this book up in the first place, and you can achieve wonders. We are here to help. Feel free to reach out anytime. Through the power of robust video conferencing that is now available, we could be meeting "face-to-face" quickly without having to plan significant travel.

The SEF program has a long and distinguished history. The generous support of our sponsor has allowed the program to flourish. This support combined with the expertise

A Roadmap for Transformative Science Teacher Leadership

and passion of dedicated faculty and staff around the United States has built a program that can help school districts deliver for their most important constituents – their teachers and their students. The real heart of the program is the more than 500 Fellows and DSCs that have dedicated thousands of hours to living this program. Without them, the program is little more than an instruction manual and an idea. We leave you with some closing thoughts of alumni Fellows.

The [SEF] Fellowship has been one of the best professional development experiences I have encountered in over a decade of teaching. While I wanted to get some quick NGSS training, I ended up being a part of something much more meaningful. In my V-CCLS group, I not only explored group work in the classroom setting, but I also had a chance to sit down and collaborate with teachers at the high school level. We were able to really look at our science teaching in a new and valuable perspective. In my H-CCLS group, we looked at inquiry in different stages and how different the implementation could go based on grade level and classroom culture. It was an eye-opening experience that generated a lot of great discussion and reflective thinking.

Angela Bisbee, CA, elementary teacher, 2019

I remember my Principal bringing me the application for the USF [SEF] Fellowship in April of 2018 just four days before the deadline, and it was Easter weekend. I remember feeling unsure of what exactly I was getting myself into, and then, I remember being very excited when I got the e-mail notifying me of my acceptance as a Tampa Bay [SEF] Fellow. The most accurate word to describe my first year as a [SEF] Fellow is validation. During my first meeting, I was placed into this unique group of wonderful women from different levels of elementary and middle school education; all of us veteran teachers; and, all of us eager to be productive members of the group. Our time together, debriefing our lessons, was the most valuable time for me. We spent time sharing our fears and doubts, of course, but most important of all, we spent time offering each other the most honest and constructive feedback that I have even received in my nine years in education. The feedback from these women validated my abilities as an educator and laid the groundwork for the confidence I would need to carry out my GPS Project.

Jacqueline Bromley, FL, 2019

If I were to sum up my experience with the [SEF] program, I would say that it helped build up my confidence as a teacher for my students and for my colleagues. The connections and opportunities of the [SEF] program has gradually pushed me into the role of a teacher-leader within my campus and district because of the projects, group evaluations, and professional development provided by the educational department from Stanford. In a short two years, I have been given opportunities to not only to learn from educators that are in the classroom, I was given the chance to share my experiences with educators from around the United States.

Dean Lorenzo, CA, 2020

Chapter 8

Conclusion and the path forward

Arthur Eisenkraft

This book and the program it describes aim to transform science education by developing educators to their fullest potential. At its heart, the Wipro Science Education Fellowship (SEF) program is about professional learning that drives sustainable district change through teacher leadership. Since 2014, the SEF program has impacted the lives of thousands of students in high-needs school districts through the efforts of hundreds of Fellows. This collaborative initiative involves teachers, science coordinators, school district administrators, and university professors and staff. The journey detailed in this book provides insights and guidance for replicating the SEF program or adapting its elements to local contexts.

As we reach the conclusion of this book, we reflect on the transformative journey the SEF program has taken us on. From its inception to the innovative adaptations that followed, SEF has significantly impacted science education across multiple districts and universities. Let's revisit the key points from each chapter, celebrate our achievements, and look forward to the bright future that lies ahead.

The foundation: Chapters 1–3

In the **Introduction**, we laid out the essence of the SEF program and its overarching goal: to improve science education by fostering teacher leadership. This chapter also posed a fundamental question: What sets the SEF program apart from other professional learning experiences? The answer lies in its comprehensive, collaborative, and transformative approach, which has proven successful across diverse educational settings.

Chapter 1: The SEF program overview

The SEF program was born from a need to improve science education through teacher leadership. We began by building a robust partnership between universities and

DOI: 10.4324/9781003490586-9

school districts, focusing on professional development and creating a supportive community for science educators. This chapter laid the foundation for understanding the importance of collaborative efforts and the power of teacher leadership in driving educational change. The chapter highlighted how the SEF program supports district transformation by building a critical mass of teacher-leaders who work alongside science coordinators, principals, and district administrators.

Chapter 2: Theoretical background and teacher leadership

In this chapter, we explored the structure of the SEF program, detailing the two-year fellowship model. Year 1 focused on improving classroom instruction through Collaborative Coaching and Learning in Science (CCLS), while Year 2 emphasized leadership development with the Growth Plan System (GPS). This structure ensured that Fellows not only enhanced their teaching practices but also grew as leaders capable of influencing their peers and districts. This chapter discussed the importance of teacher leadership in driving educational change and improving student outcomes. It provided a theoretical framework for understanding how teacher-leaders can influence their peers and foster a culture of continuous improvement within their schools and districts. We linked the elements of the SEF program to research efforts that shaped the program including insights from the teacher leadership literature as well as that from analysis of distributed leadership.

Chapter 3: The Collaborative Coaching and Learning in Science (CCLS) model

This chapter focused on the first year of the SEF program, where the emphasis is on enhancing instructional practices. Fellows engage in CCLS, working in vertical and horizontal teams to refine their teaching methods and integrate research-based strategies into their classrooms. Through classroom observations, reflective discussions, and shared learning experiences, Fellows gained valuable insights and practical strategies to improve science instruction. This model emphasized the importance of community and mutual support in professional growth. This year is crucial for building a strong foundation of instructional excellence that Fellows can later leverage as leaders.

Building on success: Chapters 4–6

Chapter 4: The Growth Plan System (GPS)

The GPS year was a pivotal phase where Fellows developed and implemented individualized growth plans aligned with district goals. By leading professional development sessions and working on specific projects, Fellows not only advanced their skills but also contributed to broader district initiatives. This chapter showcased the potential of targeted projects to drive meaningful change with both district and personal goals. This year allows Fellows to apply their instructional expertise in leadership roles, providing professional development for their colleagues and contributing to district-wide initiatives.

Chapter 5: The role of District Science Coordinators

District Science Coordinators (DSCs) played a crucial role in the success of the SEF program. This chapter highlighted how DSCs supported Fellows, facilitated professional development, and helped align the program with district objectives. Their leadership and dedication were instrumental in fostering a collaborative environment and ensuring the program's sustainability.

Chapter 6: The role of the Institution of Higher Education (IHE)

Universities served as the backbone of the SEF program, providing resources, expertise, and a central hub for activities. This chapter discussed the mutual benefits of university-school district partnerships and how these relationships evolved into deeper, more meaningful collaborations. The IHEs' commitment to excellence in science education was a driving force behind the program's success.

Innovation and sustainability: Chapter 7

Chapter 7: Innovation Phase of the Science Education Fellowship

The Innovation Phase marked the next step in the SEF journey. Each university crafted unique plans to continue supporting district transformation through teacher leadership. From expanding cohorts to introducing new models and projects, this phase demonstrated the adaptability and scalability of the SEF program. It was a testament to the ongoing commitment to improving science education even after initial funding concluded.

Key takeaways and future directions

Reflecting on the journey detailed in this book, several key themes emerge:

1   **Collaboration and community** – The success of the SEF program is rooted in its collaborative nature. By bringing together teachers, coordinators, administrators, and university staff, the program creates a supportive community that fosters professional growth and collective problem-solving.
2   **Teacher leadership** – Empowering teachers as leaders is at the core of the SEF program. By enhancing their instructional practices and leadership skills, the program enables teachers to drive positive change in their schools and districts.
3   **Sustainability and innovation** – The SEF program's impact extends beyond its initial implementation. Through the Innovation Phase, the program continues to evolve, ensuring its sustainability and adaptability to changing educational landscapes.

4 **Flexibility and adaptability** – One of the SEF program's strengths is its flexibility. Each university site and school district has tailored the program to meet their unique needs, demonstrating that the SEF model can be successfully adapted to diverse contexts.

5 **Continuous improvement** – The SEF program embodies a culture of continuous improvement. By engaging in reflective practices, leveraging research, and adapting to new challenges, the program ensures ongoing growth and development for all participants.

Celebrating our achievements

The journey through the SEF program has been one of growth, collaboration, and transformation. We've seen Fellows evolve into confident leaders who drive change in their classrooms and districts. Districts have benefited from a more cohesive and effective approach to science education, while universities have strengthened their roles as pivotal players in educational reform.

One of the most compelling aspects of the SEF program is its emphasis on teacher leadership. By empowering teachers to take on leadership roles, we've witnessed a ripple effect of positive change. Teacher-leaders inspire their colleagues, advocate for best practices, and contribute to a culture of continuous improvement. This approach ensures that the impact of the SEF program extends far beyond the individual Fellows, reaching entire schools and districts.

Looking ahead: Replicating and adapting SEF

As you embark on your journey to replicate or adapt the SEF program, we encourage you to take inspiration from the stories and strategies shared in this book. Here are some key takeaways to guide you:

1 **Start small, think big** – Begin with a set of local school districts and a cohort of 20 teachers. Use this initial success of the first year to build momentum and expand the program to more cohorts from the same districts in additional years.

2 **Engage stakeholders early** – Involve teachers, coordinators, administrators, and university staff from the outset. Their buy-in and support are crucial for the program's success.

3 **Customize to your context** – Tailor the SEF model to fit the specific needs and goals of your districts and universities. Flexibility is key to ensuring the program's relevance and effectiveness.

4 **Leverage existing resources** – Utilize existing professional development funds, university resources, and community partnerships to support the program. Look for opportunities to integrate SEF activities with ongoing initiatives.

5 **Focus on sustainability** – Plan for the long-term by seeking additional funding and resources to sustain the program beyond its initial implementation. Explore opportunities for local, state, and national grants to support your efforts.

6 **Build a community of practice** – Foster a culture of collaboration and continuous improvement by creating opportunities for participants to share their experiences, learn from one another, and work together on common goals.

7 **Measure and reflect** – Continuously evaluate the program's impact and use this data to inform your practice. Engage in regular reflection to identify successes, challenges, and areas for improvement.

8 **Build strong partnerships** – The success of the SEF program hinges on strong, collaborative partnerships between universities, school districts, and other stakeholders. Foster open communication, mutual respect, and a shared vision for educational excellence.

9 **Emphasize teacher leadership** – Invest in developing teacher-leaders who can drive change from within. Provide opportunities for professional growth, encourage reflective practice, and support teachers in taking on leadership roles.

Final thoughts

I will say this to anyone who will listen as many times as I can...the Wipro program has been the most transformative professional development program I have ever participated in (and probably ever will). – (NJ District Science Coordinator)

As we conclude this book, we look to the future with enthusiasm and optimism. The journey toward improved science education is ongoing. The SEF program provides a powerful framework for driving positive change through teacher leadership. By embracing the principles and practices outlined in this book, you can create a transformative professional learning experience that benefits teachers, students, and entire school districts.

As the great Spanish poet Antonio Machado said, "Caminante, no hay camino, se hace camino al andar" (Traveler, there is no path, the path is made by walking). Our hope is that the SEF program and this book will serve as a guide and inspiration for you as you create your own path toward higher-quality science education.

We wish you good travels on your journey. Whether you are a university faculty member, a district administrator, a science coordinator, or a teacher, your efforts are vital to the success of this endeavor. Whether you are replicating the SEF program, adapting its elements to your context, or using it as inspiration for your path, know that you are part of a larger community dedicated to excellence in science education. Together, we can make a lasting impact on science education and ensure that all students have the opportunity to achieve their full potential.

Thank you for joining us on this journey. We look forward to hearing about your successes and learning from your experiences as you implement and adapt the SEF program in your own context. May your path be filled with discovery, growth, and transformation.

Good travels, and may you continue to make the road by walking.

Appendix A
District science coordinators

Site	District	Name
CA	Campbell Union High School District	Emily Hansen
CA	Moreland School District	Destiny Ortega
CA	Mountain View Whisman School District	Ranen Bhattacharya
CA	San Francisco Unified School District	Eric Lewis
CA	San Jose Unified School District	Diane Aronson
FL	Hillsborough County Public School District	Larry Plank
FL	Hillsborough County Public School District	Shana Tirado
FL	Pasco County Schools	Lesley Kirkley
FL	Pinellas County Schools	Fawnia Shultz
MA	Boston	Janet Bowen
MA	Boston	Nicole Guttenberg
MA	Boston	Elizabeth Hadly
MA	Boston	Pam Pelletier
MA	Boston	Molly Peters
MA	Boston	Holly Rosa
MA	Braintree	Betsy Clifford
MA	Braintree	Dianne D. Rees
MA	Cambridge	Deena DePamphilis
MA	Cambridge	Janet MacNeill
MA	Cambridge	Lisa Scolaro
MA	Malden	Douglas Dias
MA	Malden	Shereen Escovitz
MA	Malden	Diane C. Perito
MA	Pembroke	Joan LaCroix
MA	Pembroke	Jonathan Shapiro
MO	Boonville R-1	Cynthia Dwyer
MO	Columbia Public Schools	Melissa Fike
MO	Columbia Public Schools	Andrew Kinslow
MO	Columbia Public Schools	Lisa Nieder
MO	Columbia Public Schools	Mike Szydlowski
MO	Community R6	Cheryl Mack
MO	Community R-VI	Jessie Mommens

Site	District	Name
MO	Diocese of Jefferson City	Emma Williams
MO	Eldon R-1	Steve Henderson
MO	Fulton 58	Chris Hubbuch
MO	Hallsville R IV	Adym Cooney
MO	Hallsville R-IV	Bethany Morris
MO	Hallsville R-IV	Ty Sides
MO	Jefferson City Public Schools	Joseph Lauchstaedt
MO	Jefferson City Public Schools	Gary Verslues
MO	Maries R-2	Alice Taylor
NJ	Clifton	Gary Frankel
NJ	Kearny	Mary Goffredo
NJ	Montclair	Alyson Wasko
NJ	Orange	Erika Hackett
NJ	Paramus	Michael Pilacik
NY	East Ramapo	Andrea Coddett
NY	East Ramapo	Karen Lent
NY	East Ramapo	Erika Sprauer
NY	New Rochelle	Peggy Younger
NY	New Rochelle	Marselle Heywood
NY	New Rochelle	Elizabeth Barrett-Zahn
NY	Port Chester	Elsy Zizolfo
NY	Port Chester	Karla Purcell
NY	Tarrytown	Jason Choi
NY	Tarrytown	Leana Peltier
NY	White Plains	Margaret Doty
NY	White Plains	Vincent Dougherty
NY	White Plains	Carmen King
TX	Cedar Hill	Jeremy Hesse
TX	DeSoto	Allen, Raisha
TX	DeSoto ISD	Danielle Moore
TX	Grand Prairie	Tamara Majors
TX	Irving	Chris Dazer
TX	Lancaster	Faith Milika

Appendix B

University support staff

Site	University	Name
CA	Stanford Graduate School of Education	Suzanne Burrows
CA	Stanford Graduate School of Education	Sylvia Cardenas
CA	Stanford Graduate School of Education	Sarah Mandudzo
CA	Stanford Graduate School of Education	Preetha Memon
CA	Stanford Graduate School of Education	Sharon Parker
CA	Stanford Graduate School of Education	Daniel Pimental
CA	Stanford Graduate School of Education	Jennifer Ray
FL	University of South Florida	Nancy Islam
FL	University of South Florida	Karl Jung
FL	University of South Florida	Katie Laux
FL	University of South Florida	Larry Plank
MA	University of Massachusetts Boston	Sarah Beberman
MA	University of Massachusetts Boston	Bob Chen
MA	University of Massachusetts Boston	Marilyn Decker
MA	University of Massachusetts Boston	Anthea Gabriel
MA	University of Massachusetts Boston	Alex Hartley
MA	University of Massachusetts Boston	Roxanne Johnson DeLear
MA	University of Massachusetts Boston	Eric Weiss
MA	University of Massachusetts Boston	Allison Scheff Little
MA	University of Massachusetts Boston	Kimberly Rocco
NJ	Montclair State University	Tim Aberle
NJ	Montclair State University	Jackie Willis
NJ	Montclair State University	Shanna Andersen
NJ	Montclair State University	Ursula Derios
NJ	Montclair State University	John O'Meara
NY	Mercy University	Kristen Napolitano
NY	Mercy University	Mary Ushay
TX	University of North Texas Dallas	Kendra L. Brown

Index

Note: Page numbers in *italics* and **bold** refer to figures and tables, respectively.

vertical collaboration 136, 143
Vertical Collaborative Coaching and Learning Science (V-CCLS): conference 44–45; course of study (CoS) 36–39; cycle 48; final documentation of teams 46, **47**; first half of the year 34–48; format of sef meeting protocol 43–44; graphical presentation of V-CCLS process 43, *44*; group 33, 44, 46, 48, 53, 101, 141, 153; live debriefing 40–41; outcomes of 47–48; preparing group portfolio 47; presentation protocol 46; presentations 10, 55; summary of V-CCLS process **43**; team 6, **35**, 36, 45, **46**; team presentation schedule 45, **46**; tuning protocol 42–43; *see also* Collaborative Coaching and Learning Science (CCLS); Horizontal Collaborative Coaching and Learning Science (H-CCLS)
vertical teams 16n1, 26, 48, 113

warm feedback 15, 40–41, 68, 91–92
website 104, 110, 111, 115–116, 143
Wenner, J. A. 23, 64
Whitworth, B. A. 20, *21*
Willenberg, Seth 76

Wipro Science Education Fellowship 1, 3–16, 17, 89, 100, 132–133, *134*, 154; CA – center to support excellence in teaching 106; Collaborative Coaching and Learning in Science Communities (CCLS) 12–13; corporate sponsorship 11; distributed leadership 8–11; findings from evaluation of 136; Individual Growth Plan Systems (GPS) 13; key components of 11–13; logic model 133, *134*; overview 3–8; supporting and incentivizing work of teacher-leaders 12; university and public school district partnership 11–12; working across partner districts 12; *see also* Science Education Fellowship (SEF)
working across grade levels to improve grades 3–5 science teaching 141–142
writing and revising GPS goals, template for 77

year 2 timeline and schedule, sample of 64, 78–80
Yichang Liu 54, 92
York-Barr, J. 23, 64